Standards and **Quality**

Anwar El-Tawil

Standards and Quality

World Scientific

NEW JERSEY • LONDON • SINGAPORE • BEIJING • SHANGHAI • HONG KONG • TAIPEI • CHENNAI

Published by

World Scientific Publishing Co. Pte. Ltd.

5 Toh Tuck Link, Singapore 596224

USA office: 27 Warren Street, Suite 401-402, Hackensack, NJ 07601

UK office: 57 Shelton Street, Covent Garden, London WC2H 9HE

Library of Congress Cataloging-in-Publication Data
El-Tawil, Anwar.
   Standards and quality / Anwar El-Tawil.
     pages cm
   Includes bibliographical references and index.
   ISBN 978-9814623575 (hardbound : alk. paper) -- ISBN 9814623571 (hardbound : alk. paper) --
   ISBN 978-9814623582 (e-book) -- ISBN 981462358X (e-book) --
   ISBN 978-9814623599 (mobile) -- ISBN 9814623598 (mobile)
   1. Standards, Engineering. I. Title.
   TA368.E43 2015
   620.002'18--dc23
                         2014025628

British Library Cataloguing-in-Publication Data
A catalogue record for this book is available from the British Library.

Typeset by Stallion Press
Email: enquiries@stallionpress.com

Printed in Singapore

DEDICATION

This book is dedicated to the memory of Dr. Lawrence D. Eicher, Secretary General of ISO from 1986 to 2002.

PREFACE

It is well known that standards and quality are closely related. While many books and references have been written on quality techniques, much less was written about the techniques of standard setting and the institutions involved in that activity at the industry, national and international levels.

This book is about standards and quality. The book provides the essential features of both disciplines. It starts with an overview of standardization: what are standards, what benefits they bring, how they evolved. Then it goes on to describe the essential features and bodies active in national and international standardization. Conformity to standards is discussed, as well as accreditation and the bodies that carry out these activities.

Quality and quality management are described. A brief history of the evolution of quality concepts from preindustrial times till the present is given. Quality techniques are described including quality management systems according to ISO 9000 series of standards and the six sigma approach to quality management.

The books also includes an overview of other management systems such as food safety management, social responsibility and energy management. The role of metrology is described and the different elements of the quality infrastructure are explained.

CONTENTS

LIST OF FIGURES

CHAPTER 1

STANDARDS AND THEIR BENEFITS

1.1 WHAT ARE STANDARDS?

In common language the word "standard" signifies several things and is used in many different contexts. The Oxford English Dictionary gives the following meanings of the word standard: 1) level of quality or attainment; 2) something used as a measure, norm or model in comparative evaluations; 3) a military or ceremonial flag; 4) a tree that grows on an erect stem of full height; 5) an upright water or gas pipe. The English word "standard" stems from an old French word that means a flag (cf. meaning no. 3 of the Oxford Dictionary above). This word indicates the position of a standard, which is that of a lead object that is followed by the troops of an army or by people in general.

In social sciences the word standard is used in the expression "standard of living", in economics we encounter the "gold standard", in animal husbandry the "breed standard", in mathematics the "standard deviation" and in cryptology the "encryption standard".

This and the following chapters deal with **documentary standards** applicable to production and service processes and to the outcome of those processes: products and services. Documentary standards are best described by the International Organization for Standardization (ISO) and the International Electro-technical Commission (IEC), the two apex international standards organizations. The definition of "standard" given by these two organizations is as follows[a]:

---

[a]This definition is given in the publication authored by the two international organizations entitled « ISO/IEC Guide 2: 2004 Standardization and related activities -- General vocabulary »

A Standard: is a *document established by consensus and approved by a recognized body that provides for common and repeated use, rules, guidelines or characteristics of activities or their results aimed at achieving the optimum degree of order in a given context.*

Several important characteristics of documentary standards follow from this definition. First, it implies that standards are established by consensus, which means that those concerned by the standard, usually called the stakeholders, have agreed that a given standard is the best way of doing business and that it safeguards their interests in a balanced manner. Secondly, the definition mentions that standards are established for common and repeated use. This distinguishes a standard from an agreement between business partners that deals with one particular situation and may be fixed in a contract or in letters of acceptance but need not be fixed in a document available to other parties such as a standard. Thirdly, the definition describes the possible content of a standard as being rules, guidelines or characteristics that apply to activities or processes as well as to their outcome. Such an outcome could be a product or a service.

The definition also states that standards are established by a recognized body. This limits the establishment of formal standards to recognized bodies that are in a position to get the stakeholders of standardization together in a way that promotes the establishment of consensus. Such bodies provide a suitable forum for precise formulation, further development and updating of the standard. Those bodies facilitate the publication of standards, make them available to users and act as a repository for the collection of standards in the given field or fields. For a more detailed description of the status and functioning of standards bodies, see Chapter 3.

An important aspect of documentary standards as defined by ISO/IEC is expressed in the last phrase of the definition, which states that a standard *aims at achieving the optimum degree of order in a given context.*

Obviously, standardization is about putting order in a given situation. Better order is ensured by specifying characteristics of the product or service. Order is also improved by limiting variety, since unlimited variety may lead to an unacceptable increase in the cost of the product or service not matched by an increase in the degree of satisfaction of the user. Searching for the optimum degree of order is one of the prime tasks of standardization. Those engaged in standardization are constantly looking for the best possible compromise between two extremes: on one hand, an excessive variety that leads to an increase in cost not matched by a worthwhile improvement of the product or service and, on the other hand, a severe limitation of variety which may lead to a reduction of satisfaction of the customer or user of the product or service. A good example of the *optimum degree of order* is evident in the case of sizes of clothes and shoes. Too many finely spaced sizes would lead to a great increase in cost as the advantages of mass production are lost and stocking and distribution become more difficult. On the other hand, too coarse a gradation of sizes would mean badly fitting clothes and uncomfortable shoes!

1.2 THE BENEFITS OF STANDARDS

Standards have numerous benefits and this fact explains their widespread application in most areas of human activity. Standards are especially important to ensure product safety and minimum quality, to facilitate compatibility between product components and between products, to optimize variety and to facilitate communication and the provision of information about products, services and situations.

1.2.1 Standards ensure product safety and minimum quality

Most products can be a source of hazard to the health and safety of people, if not properly made and packed or if not properly stored and used. Examples of such products are food products, toys, electric and gas appliances. Standards play an important role in ensuring the safety of such products by requiring that mandatory safety measures be incorporated into the products, not allowing sources of risk and specifying adequate packaging, storage and use instructions to be provided to the users.

Standards for fire-prevention and fire-fighting equipment play an equally important role in protecting people and property against fire hazard. A fire extinguisher in a building or in a car manufactured and maintained according to standards gives assurance that it will perform adequately and will not fail in the critical moment of a fire breaking out. Equally important is the written and/or graphical information on how to use the extinguisher, which should be provided in a clear, easily understandable manner.

For all products, whether hazardous or not, standards provide an important benefit as they guarantee a minimum level of quality. Standards for products and services often specify minimum use and efficiency characteristics which, if absent in the product or service, would lead to a waste of resources rather than a tangible benefit for the user. For example, if a product or appliance does not give good and efficient performance or breaks down after a very short time, this would mean a waste of consumer money and of natural resources.

1.2.2 Standards facilitate market entry and promote competition

The existence of standards facilitates the job of new producers of goods and providers of services who wish to enter the market. By consulting existing standards, those potential producers and service providers can get a reasonable idea of which characteristics should be present in the products and services they intend to offer. This allows smooth planning of production or service provision and good chances of acceptance of the resulting product or service in the intended markets. Easier market entry promotes competition as new producers can relatively easily start production or service provision. This promotes the interests of consumers since competition and the avoidance of monopoly situations help improve quality and reduce prices.

1.2.3 Standards are essential for trade facilitation

Standards are extremely useful for trade facilitation. The existence of standards makes it much easier to set-up clear contracts: rather than trying to specify in the contract the detailed characteristics of the contracted product or commodity, it is sufficient to refer to an existing standard. Not only would this save pages and pages of specifications, it would also save the possible haggling over the actual specifications themselves! In addition, the reference to a standard would reduce to a minimum possible litigation at the time of delivery: standards not only specify important characteristics of the product or service, they also specify how to sample and test the product and how to interpret test results.

1.2.4 Standard are necessary to ensure compatibility

The modern world provides numerous advantages to people many of which are based on compatibility. Compatibility exists on two levels: a) between parts and components of a product and b) between products.

When a part of a complex machine such as a car or a plane is broken or wears off, it is possible to replace it with an equivalent "spare" part produced by the same producer or by another producer on the other side of the globe, which then performs the same functions as the original part. This simple act is so commonplace in today's world that people take it for granted and are mostly unaware of the sophisticated standardization and engineering effort behind it. The **"interchangeability"** of parts and components, as this is called, is based on standardization of fitting sizes of the parts as well as their other properties such as strength, hardness, electrical properties (in the case of electric components), resistance to rust and chemicals and others. The advantages of the interchangeability of parts and components are too obvious to be enumerated. Suffice it to say that, in the absence of interchangeability, every bolt and screw would fit only into a specially made nut or screwed hole[b]; sophisticated machines would be extremely vulnerable to breakdown, as failure of

---

[b]Three of the first international standards published by the International Organization for Standardization – ISO 7-1, ISO 7-2 & ISO 68 deal with screw threads

their weakest component would lead to a breakdown and loss of value of the whole machine or would make repair extremely costly.

Compatibility between different appliances and machines makes it possible to use together different devices having different functions. For example, a digital camera using a memory card can be connected to a computer; pictures can be transferred from it to the computer for editing, storage, posting and e-mailing; the pictures can be projected on a screen using a projector or printed using a printer. In this case, **"interoperability"** of the camera, the memory card, the computer, the e-mail system, the projector, the TV and the printer is made possible by standardization of the inputs and outputs of each of these devices as well as the connections between them. The Universal Serial Bus (or USB as it is commonly known) is a marvelous example of a standardized device that enables the connection of different information technology appliances. Other examples of interoperability abound not only in information and communication technologies but in transport, construction and other fields as well.

1.2.5 Standards improve economy by reducing unnecessary variety

Variety is a costly good. Consumers and users who wish to receive products and services that fit exactly their individual needs are quite aware of this fact. As individual needs vary with the individuals themselves, an infinite variety would be needed to completely fulfill the desire to have products and services that fit exactly each individual need. This would mean individual production and/or fitting of each product and service to the needs of the particular customer. In the majority of cases, this implies a great increase in cost, as the advantages of high volume or mass production can no longer be realized. Standards help solve this problem by pointing to an optimum variety that would satisfy needs to a reasonable degree without raising the cost of the product or service to a level prohibitive for most consumers.

The theoretical basis for an optimum choice of variety was laid by Colonel Charles Renard of the French Army in the 19th century. "Preferred numbers", proposed by Renard allow an optimum choice of variety in a particular situation. Renard's preferred numbers incorporated in the International Standard ISO 3 published by the International Organization for Standardization in 1952 offer a choice of variety from a coarse gradation (low variety) to an increasingly finer gradation (high variety).

1.2.6 Standardization of graphical symbols and colors facilitates communication in all fields

International travelers appreciate the value of standardized symbols and colors which allow them, irrespective of the language or the country where they happen to be, to use cars safely without infringing traffic rules, use public transport, find their way in buildings, airports and stations (in particular, find exits, find toilets and use elevators), use equipment, find what they are looking for in health care establishments, be warned about dangers and do a host of other things. This is made possible thanks to agreements incorporated in international standards such as the ISO 7000 series — Graphical Symbols. The use of standardized graphical symbols continues to grow and this has salutary effects on communication in the fields of traffic, sports, science and many others.

1.3 THE EVOLUTION OF STANDARDIZATION

Standardization is not a new activity: it has existed in societies since the dawn of civilization. Ancient civilizations standardized weights and measures and this facilitated the replacement of barter by a monetary economy. The ancient written languages such as ancient Egyptian hieroglyphics and the Chinese script were a manifestation of the standardization of graphical symbols to convey given meanings. Ancient civilizations also standardized the calendar, which in its primitive form consisted of 12 months of 30 days each, then was corrected to include five "forgotten days" and subsequently corrected again to include the leap year and its modifications that coincide with the solar year in a more faithful manner.

Ancient civilizations applied standardization to building construction. This made possible the marvelous construction projects that mark the great civilizations of the past. In ancient Egypt, the great pyramid of Khufu was built over 20 years (around 2560 BCE) of an estimated 2,300,000 stones of standard size. In China, the Great Wall built by Emperor Qin Shi Huang starting 221 BCE followed a standardized pattern over its length of 10,000 Li (1 Li is equal to 500 meters). With 25,000 standardized watchtowers, the Great Wall of China is a good example of standardization applied to great construction works of ancient times.

Centuries after these amazing manifestations of standardization in ancient times, standards took on new dimensions with the industrial revolution. The use of new types of power such as steam and electric power exposed workers and the public to new hazards.[c] As accidents multiplied, many of them fatal, the industrializing countries responded by establishing standards for the safety of pressure vessels, gas and electric machines. The industrial revolution also saw the development of mass production and the introduction of interchangeability of parts of complex machines. This made necessary the establishment of standards to facilitate the production and supply of spare parts.

Later, standards were established to facilitate inter-operability of machines and devices. In 1904, a great fire broke out in Baltimore. When firefighters rushed from far to the city to extinguish the fire, they found that their hose couplings could not be attached to the hydrants or to other

---

[c]In 1884 alone approximately 10,000 boiler explosions and failures occurred in the US. In 1865, three of the four boilers of the Sultana, a Mississippi side-wheeler, exploded. The vessel burned to the waterline with a death toll estimated between 1200 and 1500. This is the US worst marine disaster in history.

hoses. This prompted the standardization of hose couplings by the US National Fire Protection Association.

Pipe sizes and sanitary fittings were standardized to facilitate the construction of plumbing systems in homes. The development of the automobile industry based on the assembly of parts and components produced by many small producers gave a strong impetus to standardization in mechanical engineering. In electrical engineering, plugs and sockets were standardized and so were lamps and their fittings as well as voltages and frequencies utilized in electricity production and distribution. In building construction, modularly coordinated spaces are built with standard bricks and blocks. In information technology, standardization had a dominant role from the start, which enabled the interoperability of devices and facilitated the interconnection of individual computers into a worldwide network — the Internet. The era of standardized products, services, buildings, information and communication was started and is still in full swing.

1.4 SHOULD STANDARDS BE MADE MANDATORY?

A question often debated by those concerned by standards is whether standards should be made mandatory in view of the many benefits they provide to society. The answer to this question in centrally planned economies (for example, the ex-Soviet Union and its satellites) was in the affirmative. It was argued that it is in the interest of society to make all standards mandatory and, indeed, standards were used in conjunction with quantitative planning to evaluate the performance of economic units such as factories and collective farms. This approach provided a substitute for the evaluation of the success or failure of enterprises in a situation where the traditional market indicators profit and loss played a much lesser role than in a free market economy.

In advanced market economies, however, the general attitude is that, while some standards should be made mandatory to safeguard the health and safety of people and the environment, the majority of standards should remain voluntary. The reasoning behind this attitude is that mandatory standards for all products and services would put unnecessary constraints on innovation and that market forces are sufficient to guide producers and service providers to produce products and offer services that satisfy the needs of consumers. Indeed, those innovative entrepreneurs who wish to propose a new design for existing products or a new line of products find it easier to act, where there are no mandatory standards that oblige them to follow a particular solution.

To avoid confusion in terminology, the Agreement on Technical Barriers to Trade of the World Trade Organization uses the term "standard" to designate voluntary standards and the term "technical regulation" to designate mandatory technical rules. Technical regulations can be based on existing voluntary standards that have been transformed into mandatory regulations or can be established independently. In market economies, it is customary to reserve the establishment of mandatory technical regulations to the cases where the health and safety of people and the protection of the environment are involved. For a more detailed discussion of the relationship between voluntary standards and mandatory technical regulations, see Chapter 6 – Standards and Trade.

1.5   SHOULD THERE BE STANDARDS IN ALL AREAS OF HUMAN ACTIVITY?

In literature and the creative arts the words "standard" and "standardization", have negative connotations. Standardization with its stabilizing effects could be an obstacle to innovation, which is the essence of creativity. Consequently, in spite of the advantages of standardization in the technical and economic fields, standards are not promoted in the creative arts except where they are necessary as part of the material basis of those arts.

Examples of standards related to the material basis for the fine arts are the standard frequencies of musical notes (although non-standard frequencies also have their place in modern music), the standards for paints, colors and painting supports in painting and the standards of sizes of paper and books and for typing fonts in literature.

Standards are needed and should be used to provide a reliable and permanent material basis for the arts. However, free creativity as the essence of art should not be hampered by standards. Writers, poets, musicians and artists are expected to go beyond the known and the common in search of novel and enriching experiences for the human mind.

CHAPTER 2

THE STANDARDIZATION PROCESS

2.1 ESTABLISHING NATIONAL STANDARDS

Standards can be established at different levels: company, industry, national, regional or international. This Chapter deals with the process of establishing, promoting and maintaining national standards. Establishing regional and international standards is dealt with in Chapter 4.

The process of establishing and maintaining national standards usually includes the following stages:

a) Establishing the Standards Development Plan, often referred to as the Work Programme of the national standards body
b) Developing the standards
c) Approving the standards
d) Publishing the standards
e) Promoting the standards to potential users
f) Maintaining the standards

Some of these stages may overlap. For example, a future standard may be promoted before it is finally approved and published. Maintaining standards is a continuous process that involves standards that have been already published.

2.2 THE STANDARDS DEVELOPMENT PLAN
OR WORK PROGRAMME

Most national standards bodies go about establishing national standards according to a previously set plan. Some of these bodies establish a yearly plan while others set a plan encompassing several years.

The following factors should be considered when setting the standards development plan:

a) The important sectors of the national economy. The importance of economic sectors can be judged by their contribution to the Gross Domestic Product (GDP) of the country as well as by their contribution to foreign trade and export. For many countries, exporting commodities and products has special importance, since it is the main source of foreign currencies, which can be used to finance the vital needs of the population.

b) The social needs of the population, such as the need to protect the health and safety of people and their welfare in general and the need to protect the environment. These needs become evident, for example, as accidents due to faulty products happen and result in loss of life or property or damage to the environment. Typical standards meant to protect health, safety and the environment are those for fire prevention and firefighting, food products, electric and gas appliances, pressure vessels, standards dealing with dangerous chemicals and ionizing radiation and those dealing with the transport of dangerous goods.

c) The specific needs of stakeholders of standardization in the country, who should be regularly asked to provide their standardization needs to the national standards body.

d) If the country belongs to a regional economic group, attention should be given to existing regional standards and to plans for setting such standards.

e) The resources available to the national standard body to carry out standardization work.

Investigation of the above factors may yield results at different levels. Some studies may point out to the need to give special attention to setting standards for particular economic sectors (such as the textile industry) or particular export products (such as cotton yarn). Other investigations may highlight particular subjects for standardization (such as the safety of domestic gas appliances). Combining the different types of results to obtain an optimum yearly or multi-yearly standards development plan is a complicated task facing all national standards bodies.

The International Organization for Standardization (ISO) has developed a methodology for setting national standardization strategies for the benefit of its members. This methodology proposes the use of a weighting system for combining the different estimated priorities into a single priority for each proposed subject for standardization (or work item), resulting in a multi-yearly standards development plan based on priorities and available resources.

Details of the ISO methodology are available on the Organization's website at www.iso.org.

2.3 DEVELOPING AND APPROVING STANDARDS

Once the Work Program of the national standards body has been approved, the standards development process can be started. This process consists of a number of stages.

First, there is a ***preparatory stage*** carried out usually by the staff of the national standards body. Standards engineers (as this staff will be called in the following) search for references for the future standards. References may be international standards or drafts, regional standards or drafts where they exist, national standards of other countries or other documents. The search should be carried out in this order for the following reasons.

International standards should be given priority in application of the Technical Barriers to Trade (TBT) Agreement of the World Trade Organization. This Agreement, which is mandatory for all WTO members, mandates the "use of international standards or parts of them where they exist or their completion is imminent, as a basis for national standards, except where they are ineffective or inappropriate, for instance, because of an insufficient level of protection or fundamental geographic factors or technological problems".[d] The logic for this requirement of the TBT Agreement is obvious: if all countries adopt international standards as a basis for their national standards, technical barriers to trade between countries due to differing standards would be eliminated and international trade would be greatly facilitated.

Countries that are members of regional groups are interested in aligning their standards with regional standards. For example, members of the European Union are interested in adopting European standards as national standards since this is necessary to facilitate their trade with other members of the EU. So that the second priority after international standards should be to use regional standards, if the country belongs to a regional economic group, as a reference for national standards. Countries that do not belong to a regional group but have important trade links with a particular regional group might also wish to align their national standards to the regional standards of that group.

Where no international or regional standards exist, references for national standards could be sought in the national standards of foreign countries that have strong trade relations with the country concerned. Care should be exercised at this stage not to isolate the country for a long time by aligning its standards to standards which are deemed outdated or having no future. Finally, references for national standards should be sought in standards or rules applied by important local industries.

The next stage in standards development is establishing the technical committees (TCs) that will study the subjects at hand and provide their technical opinion. It might seem that, where it was decided to adopt an international or a regional standard, there is no need to establish a technical committee. However, there is still an important role to be played by the technical committee. The committee should study the content of the international standard to see if it is possible and desirable to implement it under local conditions.

---

[d]Annex 3 to the Agreement on Technical Barriers to Trade

Technical committees should be established in such a way that they include a balanced representation of national stakeholders and include the best expertise available in the country in the area concerned. Stakeholders include manufacturers, users (including consumers), traders (including local traders, importers and exporters), test laboratories, certification bodies and technical experts. The presence of representatives of the different stakeholders, who may have conflicting interests, is the guarantee that the developed standards would be based on true consensus achieved at the national level. Since a high level of expertise is needed for standards development, the technical committee should include experts some of whom will be representatives of manufacturers and users of the product, while others should come from universities, scientific research institutes, and test and certification bodies.

Each of the above groups of stakeholders have their particular interests and characteristics that require a special approach to each group. Manufacturers usually have the financial and technical means to participate effectively in standards development. The limitation in their case is that only a limited number of them can be invited to send representatives to the technical committees, so that they would not dominate the discussion. While this might be the important issue concerning large manufacturing companies, small and medium sized enterprises (SMEs) have the opposite problem: they are interested in participating in standards developing committees, but may not always have the financial means and the expertise required for participation. They should be assisted in presenting their point of view, for example, by inviting a collective representative of their trade associations.

Another group of stakeholders that needs support to be able to present their point of view concerning standards is consumers. They are the most numerous group but also the technically weakest. One way of assuring good representation of consumers is for the national standards body to appoint a technical expert charged with liaising with consumer associations and representing them at the committee meetings.

In order to effectively follow-up international standardization work and to align national standardization to it, it is recommended to establish national committees in parallel to existing international technical committees. Such national technical committees are known as "mirror committees". If a one-to-one relationship between national and international committees cannot be established due to lack of resources or any other reason, each national technical committee should be "allocated" one or more international technical committees whose work it should follow-up.

The two most important players in the technical committee (TC) are the chairperson and the technical secretary. The chairperson should preferably be an eminent expert and neutral. University professors and research workers could easily fulfil these two conditions. The chairperson should have a good knowledge of the subjects dealt with by the technical committee and additionally possess the ability to lead the members to an agreement. An important quality required of the chairperson is that s/he is able to devote enough time to the work of the technical committee.

Work in the technical committee: During committee meetings, the chairperson should lead but not dominate discussions. S/he should tactfully stop useless discussions and talkative members. The chairperson should be capable of representing the technical committee, if necessary, inside the country and abroad. S/he should establish a fruitful relationship with the secretary of the TC.

The technical secretary of the committee is the representative of the national standards body (NSB). Consequently, s/he is the host of the TC and should be capable of representing the NSB in an adequate manner. The technical Secretary should have an excellent knowledge of the procedures for technical work of the NSB, since s/he is responsible for their application.

The Secretary prepares the meetings of the TC in consultation with the chairperson. This includes fixing the dates of the meetings, preparing the draft agenda and the meeting reports, circulating to TC members the draft report of the last meeting, the texts of the drafts under consideration, references such as international and regional standards. The secretary should compile the comments received on the drafts in such a way as to facilitate decision making during the meeting.

The TC secretary is also responsible for the follow-up of international drafts, presenting them to the TC and obtaining decisions on how the country should vote. The TC secretary is responsible for managing the development of each national standard as a project by setting target dates for the main stages, identifying problems and proposing solutions. After a draft standard has been approved, the secretary follows up its publication.

The TC secretary should have a suitable technical background to be able to understand the content of the standards for which s/he is responsible. However, the TC secretary is not expected to be an expert in all the fields s/he has to deal with. The secretary should have the capacity of good organization, be courteous and hospitable towards the committee members.

Where a technical committee has a broad field of activity, it can be subdivided into sub-committees (SCs) each with its own chairperson. The same secretary could serve the main committee and the subcommittees, or s/he could have other standards engineers to help. Technical committees and sub-committees could establish working groups with a limited number of experts to study and propose a draft standard or a set of related draft standards. These drafts should subsequently be submitted to the main TC or SC for discussion and approval.

Achieving consensus on technical questions is not always easy. The Chairperson and the Secretary should, therefore, discuss thorny issues together *before the meeting to try to prepare solutions*. It is useful in this concern to classify the points of disagreement by importance, to start by considering the major points and to think about the reasons behind the position of the different parties. It is equally important to separate technical issues from purely editorial ones.

Where consensus cannot be reached on a whole question, dividing it into parts and discussing each part separately might help. Choosing the solutions proposed by international standards should always be preferred, since they represent global consensus and reflect the contemporary state of the art. In case consensus cannot be reached during a committee meeting, the chairperson should propose the postponement of the discussion and/or the creation of an ad hoc group to study the matter. If all efforts to achieve consensus fail because of a small objecting minority, it might be necessary to adopt the solution supported by a solid majority of the members of the committee. Voting is not the optimum way to achieve consensus, but in some cases, it might be the only way out. The minority group may request that their objection and their point of view be recorded in the committee report and this wish should, of course, be granted. Obviously, consensus should not be established at the expense of the quality or the technical coherence of the standard.

Agreement reached on main issues should always be recorded in the report of the TC meeting under "Resolutions". In this way, the committee avoids reopening the debate in the following meetings, which could lead to interminable discussions. The meeting report should be short and concise, unless there is a good reason to the contrary.

Approval of draft standards: Drafts prepared by technical committees undergo a further process of approval. First, they are circulated for comments to all possible national stakeholders. Many national standards bodies post draft standards on their Websites. However, it is always desirable to inform stakeholders (for example, by a circular e-mail) that a new draft has been posted, so that interested parties could view and comment on it. After a suitable period, the comments are collated and presented to the technical committee. Comments may be of an editorial or substantive nature. Both should be considered and dealt with appropriately. A new draft may be circulated, if the substantive comments are important. The process is repeated until consensus[e] is reached. Some national standards bodies leave the decision as to whether consensus reached is satisfactory to the technical committee. Others require approval by a higher instance, which could be a sectorial board or it could be the Management Board of the institution where the different groups of stakeholders are represented. With the increasing number of standards being developed at present, this latter practice (approval by the Management Board of each standard) has become very rare and it is, indeed, not recommended.

It is important that the national standards body establish and follow well-defined procedures for the standards development process. Such procedures should define the rules for each stage of the process, the roles and responsibilities of the different players. The standards development procedures should be clear and unambiguous, so that they can be followed easily and can provide a basis for arbitration where differences appear.

2.4 PROMOTION OF THE USE OF PUBLISHED STANDARDS

The promotion of the use of published standards is a necessary activity that should be carried out by national standards bodies. Where the process of standard development was carried out properly, a large number of stakeholders would already be aware of the existence of the standard. However, many more individuals at the stakeholder level

---

[e]ISO and the IEC define consensus as « general agreement, characterized by the absence of sustained opposition by any part of the concerned interests and by a process that involves seeking to take into account the views of all parties concerned and to reconcile any conflicting arguments. » " NOTE — Consensus need not imply unanimity"

would still not be aware that the standard has been published and/or would not know its exact content in its final form.

Publicizing the use of national standards is of interest to the national standards body, since the sale of standards is in most countries an important source of revenue for the NSB. Even in countries where the NSB is generously financed by the Government, it would be a waste of public money that the standards go unnoticed by stakeholders who would individually and collectively benefit from their implementation. It should not be forgotten that coherent standards are useful not only to those who implement them directly, but also to individuals and entities that have activities related to the field of the standard. For example, standards for the sizes of paper would affect not only the paper industries, but they would also affect the publishing industries, individual users of paper and generally reduce the waste of paper, which would in turn reduce deforestation pressures.

The promotion of the use of standards begins by providing the basic information on published standards. Such information includes the title, the year of publication, the number of the standard, international or regional references on which the standard is based and a short abstract. Providing information on standards is discussed in more detail in Chapter 4.

The promotion of published standards is also a promotion of the national standards body that published them. The higher the reputation of that body, the stronger the readiness of stakeholders to contribute to the development of new standards. This is in fact a self-reinforcing process: a reputed standards body gets good contribution to its work from stakeholders, who are, in turn, more apt to implement the published standard and more ready to contribute to the development of new standards.

2.5 MAINTAINING STANDARDS

Standards like all media of technical knowledge evolve with time. As new technological processes are developed and older ones are abandoned, standards reflect the current state of scientific and technical knowledge and practice. A good example is standards for electronic components which started by standards for vacuum tubes, which were followed by standards for semiconductor discrete components (such as diodes and triodes) and, finally, by standards for integrated circuits.

National standards bodies have policies to review and revise where necessary the standards they have published. A period is set after which standards are automatically resubmitted to the technical committee that prepared them for review. The result of this review can be to confirm, revise or withdraw the standard. This does not prevent the committee deciding to start the revision of a particular standard, when it is clear that new information and practice justify such a revision.

CHAPTER 3

THE NATIONAL STANDARDS BODY

3.1 WHY A NATIONAL STANDARDS BODY?

It is generally accepted that standardization is a necessary activity in all civilized societies. Standardization at the national level often starts by professional societies setting standards they feel are necessary for regulating practice in the areas of their activity. In the United Kingdom, for example, the first standard was developed in 1901 by the *Engineering Standards Committee*, established by the Institution of Civil Engineers.

As the demand for national standards increases, the establishment of a national standards body becomes a necessity. Taking again the example of the UK, in 1929, the Engineering Standards Committee was granted a Royal Charter and in 1931, a supplemental Charter was granted changing the name, finally, to the British Standards Institution.

The national standards body plays a vital role in promoting and coordinating standardization activities in the country and in ensuring that they are carried out according to internationally accepted rules. This body also plays a vital role in associating national standardization with standardization at the regional and international levels, which is extremely important for developing trade with other countries.

In most countries the national standards body plays an important role in the assessment of the conformity of products and services to standards. The national standards body is often the pioneer in conformity assessment. Their activity in the field of conformity assessment of products, services and management systems often opens the way for private sector conformity assessment organizations.

3.2 CHARACTERISTICS OF THE NATIONAL STANDARDS BODY

The national standards body (NSB) should be a national technical body characterized by professionalism and integrity. The NSB should represent the interests of all stakeholders in a balanced manner, striving to serve the national interests without bias towards one stakeholder or group of stakeholders to the detriment of others.

Since both the Government and the private sector benefit from the activities of the NSB, they should both share in managing and financing the NSB. The NSB in a market economy should not be a technical regulatory body. The issue of mandatory technical regulations should remain in the domain of Government ministries and departments, although they will often use voluntary standards as a basis for mandatory technical regulations.

3.3 NATIONAL STANDARDIZATION POLICY

Most national standards bodies establish a *national standardization policy*. This is a document that provides principles for the standardization activity and helps guide actions in this field. The national standardization policy should be discussed by interested parties before it is approved. After its approval, the national standardization policy should be promoted widely to interested parties and to the public at large, to whom the broad outlines should be known.

The national standardization policy should clearly proclaim the following basic principles:

- National standards are based on consensus.
- The voluntary nature of national standards.

- Clear distinction between voluntary standards and mandatory technical regulations
- Commitment to adhere to the "Code of Good Practice for Standardization" of the Technical Barriers to Trade Agreement (TBT Agreement) of the World Trade Organization (WTO)
- Commitment to align national standards to international standards as far as possible

The national standardization policy could be published under this title or it could be part of the *vision and mission* of the national standards body.

3.4 STANDARDIZATION LAW

In most countries a law or similar legal instrument is issued that governs standardization activities. The standardization law usually gives the right to develop and publish national standards exclusively to the national standards body. Other organizations in the country may develop and publish standards that would not, however, have the status of national standards. The standardization law also gives the national standards body the authority to represent the country with regional and international organizations in the fields of standardization. The law may also specify certain aspects of governance, financing and controls of the national standards body.

The standardization law could also state some basic principles that apply to the elaboration and publication of standards, such as:

- The balanced representation of interested parties at the different levels of the national standards setting activity
- The wide consultation of parties interested in each standard
- The achievement of consensus as the basis for standards setting
- The voluntary nature of standards
- The desirability of harmonizing national standards with international standards as far as possible

3.5   GOVERNANCE OF THE NATIONAL STANDARDS BODY (NSB)

The highest governing instance of the national standards body, whether it is called the Governing Council, the Board of Directors or any other name should have a balanced representation of interested parties of standardization in the country.

The functions of the Governing Council of the NSB normally include:

- Establishing the policy of the NSB
- Approving the technical and administrative working procedures
- Approving the standards development plan
- Approving the establishment of technical committees and nominating their chairpersons
- Following up the implementation of the standards development plan
- Deciding the membership of the NSB in international and regional organizations in its fields of activity, such as standardization and conformity assessment
- Approving the budget and supervising the finances of the NSB
- Proposing the candidate to become the chief executive officer of the NSB and the candidates to become important directors in the organization

Where the NSB is engaged in other activities such as conformity assessment and certification, the Council would have planning and supervisory powers over those activities as well.

The governing council of the NSB may establish advisory bodies to assist it, for example:

- A Standards Development Advisory Committee and Sectorial Standards Committees
- A Conformity Assessment Committee

The chief Executive Officer (CEO) of the NSB and his staff are responsible for implementing the policies and procedures approved by the Council.

The governing council **should not** approve individual standards or be involved in the routine operations of the NSB.

3.6 DIRECTIVES FOR TECHNICAL WORK

The NSB should establish directives or procedures for carrying out the technical work. These directives should regulate aspects such as:

- Planning the development of standards and the establishment of the Work Programme of the NSB
- Establishing technical committees, sub-committees and working groups
- Nomination of chairs of technical committees and sub-committees and conveners of working groups
- The methodology of work in the technical committees, sub-committees and working groups
- Circulation of drafts and consideration of comments
- Definition and achievement of consensus
- Format of national standards
- Conditions and instances responsible for the approval of standards
- Appeals against decisions of technical committees and governing bodies of the NSB

3.7 FINANCING NATIONAL STANDARDIZATION

The largest part of the resources devoted to standardization comes from the stakeholders who serve as technical committee members as well as those who study and contribute comments to draft standards. Nevertheless, the NSB needs financing for the human and logistic support it provides to standards development and standards information, two activities that may not be financially self-sustainable.

Financing may be assured through various sources such as the sale of publications, training and information, members' subscriptions and Government support. Where the NSB is active in testing and certification, part of the revenue of these activities may be diverted to standards development, considering that it is the necessary basis of testing and certification.

In most developing countries, the NSB is a governmental body. Nevertheless, it is desirable to develop other sources of revenue since these would allow flexibility in covering expenses, permit reinvestment in the system and make the NSB generally more dynamic and market oriented.

In the establishment phase of a national standards body in developing countries, the Government should provide the initial investment to kick-start the NSB.

3.8 HUMAN RESOURCES OF THE NSB

The NSB needs to employ staff with a technical and scientific background to be able to understand standardization issues. Graduates with a university degree in engineering or science can be recruited. However, since very few universities offer undergraduate courses in standardization, the newly recruited staff should undergo some training oriented specifically toward the principles and practices of standardization.

Subjects for training of newly recruited staff members of the NSB should include:

- The history of standardization
- The components of the national quality infrastructure: standards, testing, certification, accreditation and metrology
- Principles of modern standardization

- The process of standards development
- Standards information
- Standards and trade — technical barriers to trade
- Notions of conformity assessment
- Notions of management systems
- Notions of metrology
- Language training for members who do not already have good language capabilities

Several resources may be available for training the staff of the NSB in developing countries. Examples of those are:

- Study tours with well-established NSBs abroad
- Local training using national and foreign resources (such as those provided by technical assistance agencies of foreign countries and by international organizations)
- Self-study using material of the International Organization for Standardization (ISO)
- ISO e-learning courses
- Seminars organized by the International Organization for Standardization (ISO) and other international organizations

CHAPTER 4

INTERNATIONAL STANDARDIZATION

4.1 A BRIEF HISTORY OF INTERNATIONAL STANDARDIZATION

The great empires of ancient times were, probably, the first place where the standards of one country were used in other countries under its domination. For example, under the Roman Empire, it was natural that roman currency, weights, measures and standards were used in all the countries under roman rule. The same happened during the zenith of the Spanish empire in Latin America and that of the French and British empires in their colonial dominions. This phenomenon often took place without special coercion, simply thanks to the economic and cultural influence of the dominant country. Standards in important fields such as building construction and electricity were also often copied from the dominant power, thanks to the economic influence of that power and since those standards were considered superior to the ones that existed locally before colonization.

The first attempt to propose truly international standards based on universal concepts was made when the French National Assembly, soon after the French Revolution of 1789, requested the French Academy of Science to "deduce an invariable standard for all measures and all weights". The result of this request was the establishment of the Metric System of Units, in whose conception eminent scientists of the time (including the father of modern chemistry, Antoine-Laurent Lavoisier) participated. The Metric System was internationally accepted by the countries that signed the Metre Convention in 1875, and later evolved to become the International System of Units SI (short for the French Système International d'Unités).

In the area of product standards, great progress was made when the principle of interchangeability was introduced both in Great Britain and in the USA in the beginning of the 19th century. Standardization of parts assembled into a larger object was known in antiquity (for example, in Egypt and China). It was also applied in shipbuilding in Venice, where a whole galley could be assembled on a production-line basis from standard parts, a method that enabled the Venetian Arsenale to produce one galley a day in the early sixteenth century. However, the application of that principle in mechanical engineering started much later when British arms manufacturers were able through standardization to provide spare parts for muskets sold in faraway Australia. In the USA, a great inventor, Eli Whitney, who had invented the cotton gin, proposed, "to substitute correct and effective operation of machinery for the skill of the artisan". By this, he meant to replace the individual fitting of parts of a musket by the assembly of standardized parts manufactured with sufficient accuracy so that they could be fitted together without rework. In a historically famous demonstration, Whitney demonstrated in 1801 to the President and Vice President of the USA at that time the principle of interchangeability in the assembly of muskets from parts chosen at random from a number of boxes. Another important step forward was made by Joseph Whitworth, the British engineer who proposed in 1841 a standardized thread that ended the conundrum of non-standard screws, bolts and nuts, which did not fit each other and had to be manufactured specially for each usage.

The first international organization with an important standardization activity was the International Telecommunications Union (ITU) established in 1865. This is not surprising, since the field of communications is one where harmonized standards are a prerequisite for achieving communication. The next area where international standardization was desperately needed was that of the production and distribution of electric energy. For electric energy to be distributed and used by industries and households, certain parameters needed to be standardized such as the voltage and frequency of alternating current.

This led to the establishment in 1906 of the International Electro-technical Commission (IEC).

After a number of national standards bodies were established, international standardization in fields other than electrical technology was proposed. By 1926, some 15 NSBs existed in industrialized countries. The Secretary General of the IEC led a movement supported by American and European NSBs to establish an international organization for standardization in fields other than electrical technology.

The International Standards Association (ISA) was established in 1926 and was adhered to by 15 NSBs from Europe, Japan and the USA. ISA started to develop international standards, which included standards for preferred numbers and for a system of fits and tolerances. After a period of intense activity during the 1930, ISA stopped its activity when World War II started.

After the end of the war, new initiatives appeared to restart international cooperation in standards development. In 1946 at a Conference held in London it was decided to establish a new international standards organization as a non-governmental confederation of the national standards bodies of all countries. The International Organization for Standardization (ISO) started its activities in the beginning of 1947.

Two other international organizations were later established as intergovernmental standards organizations. These are the International Organization of Legal Metrology (OIML) established in 1955 as an association of national legal metrology bodies, and the Codex Alimentarius Commission (CAC), which was established in 1963 by two UN agencies: the Food and Agriculture Organization (FAO) and the World Health Organization (WHO).

4.2 THE INTERNATIONAL ORGANIZATION
FOR STANDARDIZATION (ISO)

In 1946, 65 delegates from 25 countries met In London to discuss the future of international standardization. In 1947, the International

Figure 1 – Founders of ISO, London 1946

Organization for Standardization (ISO) officially came into existence with 67 technical committees. An increasing number of members including many developing countries quickly joined ISO. Those countries were at that time starting the process of industrial development and had not participated in international standardization before.

At present, ISO is the world's largest developer of voluntary international standards. With nearly 20,000 standards published in all fields except electrical technology, and a membership of 164 national standards bodies from so many countries, ISO standards have a considerable and growing impact on the economy of most countries.

4.2.1 ISO Membership

ISO has one member per country from 164 countries. There are three categories of members: full members, correspondent members and subscriber members. Each membership category gives certain rights and involves certain obligations.

Full members of ISO (114) include all industrialized countries and emerging economies, as well as many developing countries. Full members can participate and vote in all policy-setting organs and in any ISO technical Committee. Full members **influence** ISO standards development and strategy. They have the right to sell and adopt ISO International Standards nationally.

Correspondent members (46) are mainly members from developing or small countries. An ISO correspondent member can **observe** the development of ISO strategy by attending ISO policy meetings without the right to vote. Correspondent members can also **observe** ISO technical work and can **fully participate** in a limited number of technical committees (up to 5) of their choice. They can attend, submit comments and vote on drafts in the chosen TCs. As for the remaining ISO Technical Committees, Sub-committees and Working Groups, a correspondent member may nominate delegates and participate as an observer in any of them, but they have no right to comment or vote except in the five chosen committees. Correspondent members do not vote on international drafts outside the chosen five technical committees. They can sell and adopt ISO International Standards nationally.

Subscriber members (4 at present) are members from small and very small countries who have limited rights to observe and participate in ISO work. Subscriber members do not participate in setting ISO policy, although they may attend ISO General Assembly as observers. They can attend, submit comments and vote on drafts in not more than five committees of their choice. Subscriber members may not adopt or sell ISO standards.

A few UN members remain outside the ISO fold. However, ISO member countries generate some 98% of world gross national income (GNI) and represent around 97% of the world's population. A full list of ISO members is attached at Annex 1.

4.2.2 ISO Technical Work

ISO's technical work is mainly to develop international standards. These are developed by groups of experts delegated by ISO members who work in technical committees, sub-committees and working groups. Drafts are then commented and voted on by ISO member bodies.

ISO has established to date 288 Technical Committees (TCs) of which 235 are active at present while 53 are on standby. A technical committee is created on proposal by a member body supported by five other full members who commit to contributing actively to the work of the committee as participating (P) members. The TCs are denoted by numbers, following the order in which they were established. For example, TC 1 focusing on screw threads was created in 1947 and TC 288 on educational management systems was created in 2013. A TC on standby is one that has no work items in progress or foreseen but that is required to review the ISO International Standards for which it is responsible.

Some of ISO TCs are known as **Project Committees.** These are established when there is a need for an international standard on a specific topic that does not fall into the scope of an existing TC. Project Committees are disbanded once the standard has been published.

ISO has two joint TCs with the International Electro-technical Commission (IEC) one of which deals with information technology (ISO/IEC JTC 1) while the other, a project committee (ISO/IEC JPC 2), deals with energy efficiency and renewable energy sources. With more than 2650 published standards in its portfolio and nearly 600 items on its work program, ISO/IEC JTC1 is the most prolific of international technical committees.

Standards published by ISO in the first period after its establishment dealt mainly with hardware. Starting in the 1980s, ISO ventured more into software standards with standards on management systems constituting an important new trend. Examples of some famous and extensively used international standards outside the field of information and communication technologies are those for quality management, environmental management, food safety management, management of social responsibility, energy management, risk management, country codes, currency codes, language codes and information security management.

ISO has also published a number of Guides dealing with aspects of standards writing, conformity assessment, standards information, reference materials and others. Guides are normally prepared by Policy Development Committees of the organization. Some of these guides were prepared and published jointly by ISO and the IEC.

ISO's technical work is carried out in a decentralized manner. The secretariats of ISO technical committees are held in most cases by those ISO members who proposed their creation. Usually, when a new TC is established, its chair and technical secretary are both proposed by the member body who made the proposal to establish the TC. These proposals have to be approved by ISO Technical Management Board.

The secretary of a TC in consultation with the chair is responsible for proposing the work program, the meetings schedule, preparing the meetings, circulating drafts, collecting comments and preparing the texts of draft international standards, which are then sent to the ISO Central Secretariat in Geneva to be circulated to members for voting. TC meetings are held in all countries based on an invitation by the ISO member from that country, who is responsible for providing the meeting venue and facilitating the meeting in general. However, the inviting country is not responsible for financing the travel or accommodation of delegates to the meeting.

Stakeholders of standardization in countries, members of ISO, can participate in ISO technical work through their NSB. It is up to the NSB to ensure a balanced representation of the different categories of national stakeholders (manufacturers, users, consumers and technical regulators) in ISO technical work. Additionally, representatives of international organizations interested in the subject of standardization of an ISO TC may participate in its work as Liaison (L) members.

ISO has published to date some 20,000 standard documents in all fields except electrical technology and electronics. Most of these documents are international standards. However, when full consensus on a subject cannot be achieved at a particular point in time, the technical committee may decide to publish a *Technical Specification* (TS), which could serve as a prospective standard for provisional application. Technical specifications are reviewed not later than 3 years (renewable once) after their publication, and either converted into an international standard or withdrawn. It is also possible, prior to the publication of an international standard, to publish a *Publicly Available Specification* (PAS). This is an intermediate specification published prior to the publication of a full international standard. It may remain for 3 years (renewable once) after which it should be revised to become another type of normative document or withdrawn.

Some documents published by ISO are known as Technical Reports (TR), since they contain data of a different kind from that which is normally published as an international standard. For example, a technical report may include data obtained from a survey, information on work in other international organizations or on the state of the art in relation to national standards. Technical reports have an informative nature and should not contain matter implying that they are normative. Technical reports are regularly reviewed by the responsible committee to ensure that they remain valid. When the validity or relevance of a technical report are no longer true, they are withdrawn by a decision of the responsible technical committee or sub-committee.

ISO's technical work, including the work of technical committees, is supervised by the Technical Management Board (see 4.2.3 below).

4.2.3 ISO Governance Structure (Fig 2)

The highest instance in the ISO governance structure is the General Assembly (GA). This is a meeting of all ISO members, held once a year. During its session, the GA holds the powers of the Organization and approves its overall policies. The GA elects or appoints the ISO President, Vice-Presidents and members of ISO Council.

During the time between GA meetings, the Organization is governed by the Council, which meets several times a year. ISO Council sets policy and resolves conflicts of a policy and technical nature not resolved at lower levels. ISO Council consists of twenty members of which a number of permanent members represent the larger industrialized countries and a number of rotating elected members are from other countries from the different regions of the world.

The ISO Technical Management Board (TMB) defines and oversees the technical work. The TMB decides on the creation of new TCs and the allocation of work to them, resolves problems of a technical nature. The TMB reports to the ISO Council. The TMB consists of a chair and 14 members elected or appointed by Council.

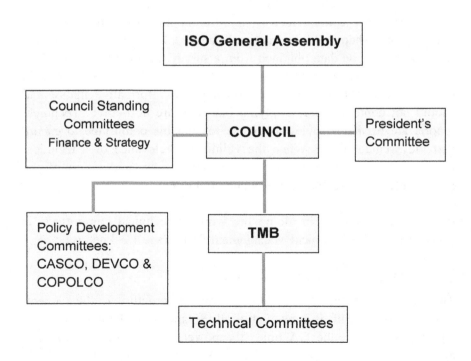

Figure 2 – ISO Governance Structure

4.2.4 ISO Policy Development Committees

ISO has established a number of Policy Development Committees (PDCs) that assist ISO Council and Technical Management Board in their work. The most important PDCs under Council are:

Conformity Assessment Committee (CASCO) – which provides guidance on the assessment of conformity to standards

Consumer Policy Committee (COPOLCO) – which provides guidance on consumer issues

Committee on Developing Country Matters (DEVCO) – which provides guidance on matters, related to developing countries.

The chairs of the Council PDCs attend the meetings of Council but have no voting powers.

The **Reference Materials Committee (REMCO)** is a policy development committee under the ISO Technical management Board that deals with standard reference materials used to check the performance of analysis and test equipment.

4.3 THE INTERNATIONAL ELECTROTECHNICAL COMMISSION (IEC)

The IEC was the first international non-governmental organization devoted to standardization in the electro-technical field. The decision to establish the IEC was made by a Conference on electro-technical standardization held in St. Louis, USA in 1904. The Committee on Standardization of Machinery of that Conference proposed "the appointment of a representative Commission to consider the question of standardization …" The Committee hoped that "if the recommendation is adopted, the Commission may eventually become a permanent one".

The IEC started its activity in 1906. The first President of the IEC was the renowned British scientist, Lord Kelvin. Work started on the standardization of terminology for electrical machinery, symbols for quantities and units, definitions of hydraulic turbines and recommendations for rotating machinery and turbines.

At present, the IEC has 163 members from so many countries. IEC members are known as the National IEC Committees. Out of these, 59 are full members and 21 are Associate Members. In addition, the IEC has established an Affiliate Member Program, subscribed to by 83 members from small and developing countries.

Full members of the IEC are authorized to participate and vote on Council documents as well as to participate in all IEC technical committees and sub-committees. They fully participate in and influence international standardization in the electro-technical field.

Associate members may not submit comments on Council documents. However, they may participate and vote in up to 4 selected TC/SCs. They are authorized to comment on the documents of other TC/SC without the right to vote.

Affiliate members may not vote. However, they may submit comments on the documents of up to 10 selected technical committees and sub-committees.

4.3.1 IEC Structure and Governance

The IEC is governed by a Council whose members are the Presidents of the National IEC Full Member Committees. A Council Board of fifteen members is derived from Council, which, together with IEC Officers, forms an Executive Committee.

The standardization work of the IEC is carried out by 175 Technical Committees and Sub-committees supervised by the Standardization Management Board (SMB). Industry Sector Boards and Technical Advisory Committees assist the Standardization Management Board in its supervisory work.

4.3.2 IEC Standardization Work

Standardization work of the IEC has to date resulted in the publication of some 6300 standards and more than 325 Technical Reports. IEC standards deal with the specifications and safety of electric machines and prime movers, home appliances, medical equipment, audio and video equipment and many other subjects in which electricity is the main component. In recent years, the IEC started work in several areas important for sustainability such as renewable energies (wind, photovoltaics etc.), smart electrification and electric vehicles.

4.4 CODEX ALIMENTARIUS COMMISSION (CAC)

Over the last century, the amount of food traded internationally has grown exponentially. It is estimated that international food trade amounts to 200 billion dollar a year, with billions of tonnes of food produced, marketed and transported. To respond to the growing concern of consumers about imported food safety and to satisfy the pressing need for international standards for food, efforts were made to harmonize food standards internationally. This started first at the regional level when the Council of the Codex Alimentarius Europaeus was created in 1958. This Council proposed to the World Health Organization (WHO) that it should associate itself with the Council. WHO referred the matter to the Food and Agriculture Organization (FAO) of the UN for discussion. A Joint FAO/WHO Food Standards Conference was convened in Geneva in 1962 and established the framework for cooperation between the two agencies. The Codex Alimentarius Commission was to be the body responsible for implementing the Joint FAO/WHO Food Standards Programme. After approval of this proposal by the Conference of FAO and the World Health Assembly, the Codex Alimentarius Commission was established and held its first session in Rome in October 1963.

To date, the Codex Alimentarius Commission (CAC) has 186 members from 185 countries and the EU. Any member nation or associate member of FAO or WHO which is not a member of the CAC, may, upon request, attend sessions of the Commission and of its subsidiary bodies and ad hoc meetings as observers. The Codex Commission counts 220 Codex Observers – of these 50 are intergovernmental organizations, 154 are non-governmental organizations, while 16 are UN bodies.

4.4.1 CAC Standardization Work

To date the CAC has developed and published some 470 standard documents. Of these 314 are Codex Alimentarius Standards (CODEX/STAN), 83 are Guidelines (CAC/GL), 70 are Recommended Codes of Practice (CAC/RCP) and 3 are Maximum Residue Limits (CAC/MRL).

Codex standards and guidelines are prepared by ten general subject committees and six commodity committees. Codex general standards deal with food hygiene, contaminants in food, food additives, food labelling, methods of analysis and sampling, nutrition and special dietary foods, pesticide residues and residues of veterinary drugs. Codex standards for food products deal with fish and fishery products, fresh fruit and vegetables, processed vegetables and fruits, fats and oils, sugars and spices and culinary herbs.

All CODEX standard documents are freely available on the website of the Commission.

4.5 THE INTERNATIONAL ORGANIZATION
OF LEGAL METROLOGY (OIML)

Legal metrology comprises activities for which legally binding requirements are prescribed for measurement units, measuring instruments and methods of measurements. Verifying these activities is normally performed by or on behalf of government authorities in order to ensure an appropriate level of credibility of measurement results for trade, health and safety, protection of the environment and law enforcement.

Legal metrology has numerous benefits. For the economy, legal metrology contributes to fair trade, reduces disputes, protects trading partners who have neither the skill nor the facilities to perform their own measurements. Legal metrology supports human safety by ensuring reliable measurement of parameters such as pressure, speed of vehicles on roads and maximum load of equipment. In the area of health, legal metrology allows medical practitioners to rely on their medical instruments and laboratories, making diagnosis more reliable and medical treatment such as surgery and radiotherapy more efficient and secure. Legal metrology supports environmental protection policies by ensuring reliable measurement of environmental parameters. In all these areas legal metrology sets binding rules for measuring instruments and measurements which should be complied with by law.

The International Organization of Legal Metrology (OIML) was set up in 1955 by an International Treaty to develop international cooperation in this field. The OIML publishes International Recommendations, which serve as model technical regulations for the members. It also publishes Documents and Guides. The OIML provides invaluable support to national regulators who need to draw up national regulations in fields where measurements are required.

The Organization allows sharing of experience between countries thanks to the Certificate System set up in 1991. This System makes it easier for manufacturers to obtain national type approval of a legal measurement instrument in one member based on tests carried out by other members. The Certificate System has more than 2020 registered certificates to date.

The OIML has 85 member states and 56 Corresponding members. It has published to date some 150 International Recommendations. OIML Recommendations were prepared by 18 technical committees and 47 sub-committees. OIML has also published 28 Documents and 2 Vocabularies. OIML Documents include: Principles of Metrological Control, Guidance on Training of Legal Metrology Personnel and a Model Metrology Law.

4.6 "PRIVATE" INTERNATIONAL STANDARDS

Some regional and national organizations set standards for different purposes:

- To guarantee an acceptable level of quality for their members (retailers, consumers)
- To promote the protection of the environment
- To promote socially responsible behavior of suppliers and a fair deal to the socially weak

Examples of such private standards are standards of the British Retail Consortium (BRC) and Global Good Agricultural Practice (GLOBAL G. A. P.) standards.

Gobal GAP was developed by the Northern European supermarket sector to promote good farming practices that:

- Respect the environment,
- Ensure food safety and
- Ensure safety of farming workers

GLOBAL GAP can be subscribed to by individual producers, groups of producers or by packagers.

Global GAP is a single integrated standard with modular applications for different product groups ranging from plant and livestock to plant propagation materials and compound feed manufacturing. It provides a tool kit, which allows each partner in the supply chain to position themselves in the global market.

The system documentation is organized into five major blocks (Figure 3) each with a set of complementary elements. Users can select the applicable elements of each block to create a tailor-made manual.

The Global GAP certification includes stages of certification to:

- General Rules (GR)
- Global GAP requirements

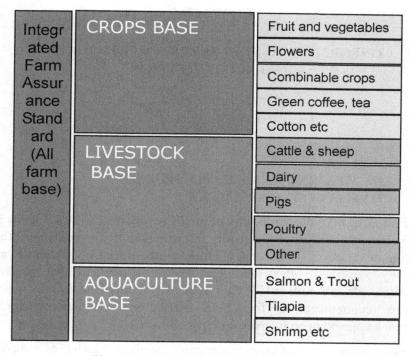

Figure 3 – Global GAP Standards Groups

- Inspection documents or checklists
- National GAP requirements referred to as Approved National Interpretation Guidelines
- Benchmarking Checklist

The Global GAP secretariat is headquartered in Cologne, Germany.

CHAPTER 5

CONFORMITY TO STANDARDS — CERTIFICATION AND ACCREDITATION

5.1 THE MEANING AND IMPORTANCE OF CONFORMITY ASSESSMENT

The assessment of the conformity of products and services to standards is an important function of the quality infrastructure that complements the setting of standards. In general, conformity assessment is defined as "an activity concerned with determining directly or indirectly that relevant requirements are fulfilled".[6] This general definition applies to cases where "requirements" are fixed by a standard or by another contractual document agreed upon by trade partners.

In most situations, conformity assessment involves drawing of samples and their testing. Normally, conformity assessment is accompanied by some form of certification or marking which could apply to a particular lot of products or to all products resulting from a manufacturing process that has been verified and approved. Such certificates and marks are often a requirement for access to markets where mandatory regulations are in vigour.

5.2 TYPES OF CONFORMITY ASSESSMENT

In any market situation, at least two parties are involved: the supplier and the buyer. In contracts and similar legal documents, these two parties are usually referred to as the **first party** (the supplier) and the **second party** (the buyer). Either of these market players can perform conformity assessment of the exchanged product to standards or other agreed requirements (Figure 4).

---

[6]ISO/IEC Guide 2: 2004 Standardization and related activity–General Vocabulary.

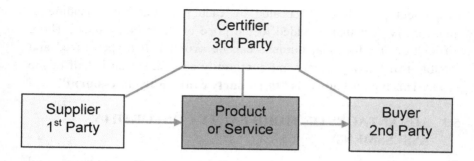

Figure 4 – Types of Conformity Assessment

The supplier, who usually carries out some testing of the product in the factory (or, if he is an importer, when approving delivery of an imported product) may simply declare that the product is conforming to a particular standard or other requirements. This declaration is known as a **"supplier's or first party declaration"**. A supplier's declaration may be acceptable to the buyer based on his confidence in the supplier and the good reputation of the latter. However, a certain amount of risk is involved in the acceptance of a supplier's declaration, specially, in view of the economic pressures experienced by suppliers to sell their products at any cost.

To avoid this risk, the buyer may wish to carry out her/his own assessment of conformity of the product s/he is buying. This type is known as a **"buyer's or second party conformity assessment"**. This type of conformity assessment is commonly practiced by corporate buyers who have the financial and technical means to carry out their own assessment of conformity of materials and components they buy. For practical reasons, the non-professional buyer or consumer does not have the capability to carry out such conformity assessment with the exception of visual and rudimentary inspection of the product.

The problem with buyer's conformity assessment is that it may be difficult and costly, in particular, if it is carried out by a staff member of the buyer at a distant place or in a foreign country. In addition, buyer's conformity assessment is usually repeated by each buyer in conjunction with their own purchase. This means that the same conformity assessment procedure is repeated by each potential buyer leading to unnecessary cost that is added to the price of the product itself. These difficulties and financial burdens can be avoided, if a specialized and reliable **third party** carries out conformity assessment on behalf of the buyer. This type is known as **"third party conformity assessment"**.

5.3 ADVANTAGES OF THIRD PARTY CONFORMITY ASSESSMENT

Third party conformity assessment provides important advantages to all market players. A supplier that obtains third party conformity assessment of his products, gets wider market access and a clear advantage in comparison with other suppliers who do not have third party approval. In addition, the supplier gets assurance that his quality management activity is well organized and is producing the desired effect. In case of the supplier being involved in product liability litigation, he can claim that he has exercised due diligence as witnessed by third party conformity assessment.

For corporate buyers who purchase materials and components, third party conformity assessment gives assurance of a higher quality of their final product and lower risk of disruption of their operations, due to assembly of non-conforming components purchased from outside the company. These advantages are obtained faster and at a cost, which is often lower than the cost of their own sampling and testing. This is particularly the case, if the sampling and testing have to be carried out at the supplier's premises in a foreign country.

As for the individual, non-professional, buyer-consumer, third party conformity assessment, as witnessed by a mark of conformity, is in most cases the only means s/he has to obtain assurance about the quality, safety and longevity of the product. It is worth mentioning, that this proof of conformity is provided at a reasonable cost (that of the quality mark) which does not add substantially to the cost of the product.

The advantages of third party conformity assessment for the national and global economies are considerable as well. For the national economy third party conformity assessment promotes consumer protection, the protection of the environment and improves public purchases. Widely practiced third party conformity assessment promotes the national economy as it reduces conformity assessment costs all over the economy, while assuring good quality. For similar reasons, third party conformity assessment facilitates international trade, contributes to the protection of the global environment and helps enhance the global economy.

The main condition for third party conformity assessment to provide the advantages mentioned above is that **it should be credible** in the target markets locally and abroad.

5.4 FUNCTIONS OF THIRD PARTY CONFORMITY ASSESSMENT

Third party conformity assessment includes a number of functions. The first is the determination of the key properties of the object of conformity assessment. In the case of a product or service, this is carried out by testing it; in the case of a management system by performing a system audit and in the case of a person by examining that person. The next function is the comparison of the properties as determined with the specifications for those properties as set by a standard or other type of requirement (such as contractual requirements). The outcome of such a comparison is to find that the requirements have been fulfilled or not. In case of fulfilment of requirements, this fact may be attested by issuing for the benefit of the buyer a certificate for the lot of products concerned. In cases, where conformity to requirements is a condition for putting product on the market, the authority responsible for market surveillance may issue a license to that effect. Many conformity assessment bodies offer a conformity (or quality) marking system. Such systems include, in addition to product conformity assessment, some form of assessment of the manufacturing process and the quality management system implemented by the manufacturer. This is used as a basis for granting a general conformity mark that can be affixed on every unit of the product and which testifies to its conformity to a standard or other requirements.

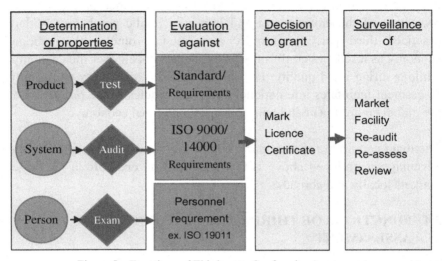

Figure 5 – Functions of Third-party Conformity Assessment

Sampling, testing and granting a certificate of conformity to a particular **lot of product** is a complete process that ends at this point. On the other hand, granting **a product that is continuously produced** a mark of conformity is a long-term process that involves the responsibility of the conformity assessment body for as long as the product continues to carry the conformity mark. Conformity assessment bodies that grant a product mark usually continue to follow the product after it leaves the factory by taking samples from the market and by re-assessing the product periodically.

In the case of certification of a management system, the certified system is re-audited periodically to ascertain that it continues to be in conformity to the relevant standard or to other requirements.

A similar situation exists in some cases of personnel certification, where the person receiving the certificate is required to take refresher courses and to pass periodic tests to ensure that s/he is still capable of performing the functions expected form them.

5.5 CONFORMITY ASSESSMENT BODIES

Different types of conformity assessment bodies exist in practice that carry out different functions. **Test laboratories** are considered the archetypal conformity assessment body which receives specimens drawn from a lot of products or a production process, tests them and issues a test report. A laboratory test report may simply give the results of the tests and analyses carried out by the laboratory or it may go on to compare those results with the requirements of a standard and conclude the conformity (or lack of it) of the sample.

Product certification schemes are usually established by national standards bodies or by specialized conformity assessment bodies. These schemes testify to the conformity of a product or line of products manufactured by an enterprise by granting a *conformity mark* affixed on the products. The conformity assessment process leading to the granting of the mark usually includes testing of samples of the product drawn from the factory and from the market, comparing the test results to a standard, auditing the quality management system implemented by the company, granting permission to affix the mark on the products and surveillance of the continued conformance of the product through further testing of samples and ensuring that the enterprise quality management system remains effective. Bodies operating product certification schemes may have their own test laboratories, or may use the services of external test laboratories.

Product certifiers may also carry out on demand conformity assessment of lots of products or do inspection work. However, specialized **inspection bodies** exist that engage in inspection and certification of products bought by companies or by the Government and other conformity assessment work such as checking the conformity of installations to fire and explosion safety criteria or other types of safety criteria. Inspection bodies play an important role in cases where products and components are bought in foreign countries, as they have their offices in many exporting countries and can act quickly and efficiently on behalf of buyers who do not have to travel to the country where the product is manufactured.

Since the implementation of management systems starting with quality management systems took off in a spectacular way in the 1980's, a large number of **management system certifiers** has been established. These bodies found a profitable niche in the area of management system certification, which, on one hand, does not require a large initial investment and, on the other hand, is a fast growing market. Witness to the size of that market is the fact that the number of companies and organizations certified to ISO 9001 standard on quality management alone has surpassed one million organizations in 170 countries. Other management system certificates should be added to this number as the number of international management system standards (such as environmental, social responsibility, food safety, energy and risk management systems) continues to grow.

Another important type of conformity assessment bodies are **personnel certifiers**. Strictly speaking, all professional educational institutions are personnel certifiers, as they give out a certificate in the form of a diploma of the competence of their graduates. Often, the diploma obtained from an educational institution should be complemented by a certificate from some professional association, before a person is admitted to practice a profession. In the technical field, examples of personnel certification are encountered in welder certification bodies, certifiers of non-destructive-testing professionals, auditor certification bodies and others.

5.6   REQUIREMENTS OF CONFORMITY ASSESSMENT BODIES (CABs)

All conformity assessment bodies (CABs) should fulfil certain requirements to obtain the confidence of users of their services and in many cases to be admitted to practice conformity assessment in the mandatory domain. ISO and IEC have developed a number of standards for the different types of CABs. For example:

- ISO/IEC 17020 – for inspection bodies
- ISO/IEC 17021 – for management system certifiers

- ISO/IEC 17024 – for bodies operating certification of persons
- ISO/IEC 17025 – for test and calibration laboratories
- ISO/IEC 17065 – for bodies certifying products, processes and services

All of these standards have common requirements, which can be resumed as follows.

5.6.1 Organization and accountability

The conformity assessment body should be a legal entity responsible for its decisions. Moreover, the final decision making should be in the hands of a clearly identified committee or person. In case of a committee, balanced representation of the different interested parties should be assured within the committee. Since a conformity assessment body may be held responsible for its decisions (such as declaring a product or an installation safe), the CAB should be financially stable and it should make liability arrangements to respond to claims for damages, in case its decisions are considered to have caused material damage or damage to the health of persons.

One requirement of international standards that helps conformity assessment bodies make sound decisions is the separation of the assessment activity (for example, the audit or inspection) from the final decision making which should be in the hands of a separate person or committee. This separation in the decision making process reduces the chances of a biased or incorrect decision as it subjects each decision to scrutiny by the committee.

5.6.2 Impartiality

As most people would agree, impartiality is one of the basic requirements of any conformity assessment body. Those bodies should be completely neutral vis-à-vis market players who use their services. Without impartiality, confidence in the judgments of the CAB cannot be established.

The policies and principles of CABs should be developed by representatives of the interested parties in a balanced manner. The criteria for the conformity assessment process should be developed by impartial committees. Those criteria should be transparent.

Impartiality should also be demonstrated on the level of accessibility of services. CABs should provide free access to their services not conditioned by size or membership of particular associations. There should be no prior limits on the numbers of entities that can apply for assessment and no undue financial burden or other conditions for accessing the services of the CAB.

Conformity assessment bodies should avoid conflict of interest by not offering:

- the same services that they assess
- consulting services to design or implement systems that they assess
- consulting services to obtain or to maintain conformity

5.6.3 Personnel of Conformity Assessment Bodies

The personnel of conformity assessment bodies should be carefully chosen by a selection process to meet the qualification criteria and personal attributes set by the relevant international standard. The CAB must keep records of the personnel engaged in conformity assessment.

Personnel should be chosen for assignments based on:

- knowledge of assessment methods
- technical competence and
- freedom from conflict of interest

The conformity assessment body should provide assessment personnel with procedures and instructions.

5.6.4 Quality Management Systems of Conformity Assessment Bodies

All international standards for CABs require that they have a quality management system with the usual elements required by ISO 9001. These elements include:

- A quality policy and quality objectives
- Appointment of a management representative for quality matters with defined responsibility to:
  - establish, implement and maintain the quality system
  - report on the performance of the quality system to top management
- A documented quality system
- Regular internal audits
- Management reviews

5.6.5 Technical Aspects of Conformity Assessment

The technical aspects of conformity assessment depend on the type of conformity assessment work (testing, auditing, inspection, certification of products, certification of management systems etc.). Requirements for those technical aspects are set by each of the relevant international standards.

5.6.6 Appeals to Conformity Assessment Bodies

The right to appeal against the decisions of conformity assessment bodies is upheld by all international standards dealing with those bodies. The CAB should look into the appeal and make an appropriate decision. Records should be kept by the CAB of both the appeal and the action taken relative to it. Those records should be made available to the authority that supervises the CAB, for example, the national accreditation body.

5.7 ACCEPTANCE OF THIRD PARTY CONFORMITY ASSESSMENT

The advantages of third-party certification can be achieved only, if certificates are accepted in target markets. In a market economy certification is offered by certifiers of different size, level and reputation.

This means that acceptance is not always assured. Acceptance of third-party certificates needs to be strengthened locally and, in today's globalized markets, internationally.

5.7.1 Acceptance of certificates at the national level

Acceptance of certificates and marks at the national level can be enhanced in two ways: 1) thanks to the good reputation of the certifying body or 2) through accreditation. Accreditation bodies have been established in most countries to check the performance of conformity assessment bodies such as test laboratories, product certifiers, management system certifiers and personnel certifiers and to confirm their competence (Figure 6). Due to the variety of conformity assessment bodies, accreditation bodies may be specialized or may be general with a number of sections each dealing with a particular type of conformity assessment body, such as laboratories, product and management system certifiers.

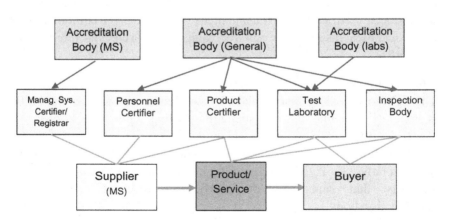

Figure 6 – Position of Accreditation Bodies

The accreditation process starts when a conformity assessment body (CAB) requests to be accredited by the national accreditation body. The accreditation body carries out an office evaluation of the conformity assessment body based on documents presented by the latter and, if it is found that the basic information of the CAB is in line with the relevant international standard, a field evaluation is carried out by evaluators of the accreditation body at the site of the CAB. In case the CAB carries out assessments at the site of its clients, such as assessments of the quality management system of an organization, the accreditation evaluators may request to witness such assessments as observers. Such witnessing would allow the accreditation evaluators to observe the manner of carrying out the assessment. The evaluators use this occasion to evaluate the thoroughness and the skill in obtaining reliable information about the client of conformity assessment. This helps the accreditation evaluators reach a correct conclusion about the competence of the CAB.

5.7.2 Cross-border acceptance of conformity assessment results

Several ways exist for securing acceptance of the results of conformity assessment in foreign countries:

- Through bilateral mutual recognition agreements between certifiers
- Through multilateral mutual recognition agreements between certifiers
- Thanks to mutual recognition agreements between national accreditors

The first approach, through a bilateral mutual recognition agreement (bilateral MRA) between two national certifiers, is suitable for two trade partners whose main trade is between themselves. To conclude such an MRA, visits are paid by representatives of each certifier to their counterparts in the other certification body to discuss procedures applied by the other certifier and to witness how conformity assessment is carried out by them. When each of the two certifiers is convinced, that the procedures and checks carried out by the other are equivalent to their own, the MRA is concluded and trade is carried out between the two countries on the basis of one conformity assessment carried out by one of the two national certifiers and recognized by the other certifier as equivalent to their own conformity assessment.

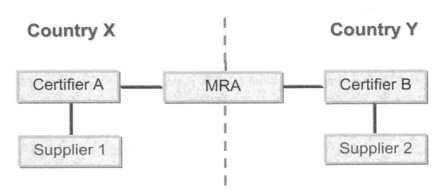

Figure 7 – Bilateral Mutual Recognition Agreement (MRA)

The second approach, through a multilateral mutual recognition agreement (multilateral MRA), is suitable for a group of countries that have extensive trade exchange between themselves, such as regional trade groups. In this case, an exchange of visits and discussions takes place between CABs of the countries belonging to the group to establish confidence in the equivalence of the conformity assessment activities practiced in the different members of the group. When such confidence has been established, the multilateral MRA can be concluded between a number of CABs from the countries of the group (one per country). This would greatly facilitate trade within the group of countries. At one point in time, this type of multilateral MRAs was popular and they played an important role in the facilitation of regional trade. However, CABs that were not members of the signing group of CABs complained that they were left out of the arrangement and, consequently, lost parts of the market.

The spread of accreditation activities to most countries have made **the third approach, namely, the recognition of conformity assessment work carried out by accredited CABs,** more common. Under this approach (Figure 8), a group of accreditation bodies of several countries mutually evaluate each other and, when they are convinced that their accreditation of conformity assessment bodies in their respective countries is equivalent, they conclude a MRA between accreditation bodies. Such an agreement signifies that the accreditation work carried out by all of the accreditation bodies is equivalent which, in turn, means that the certificates issued by any of the CABs accredited by one of the accreditation bodies is acceptable in all countries of the group.

Mutual Recognition Arrangements between accreditation bodies exist in the framework of two international organizations. The International Laboratory Accreditation Cooperation (ILAC), an association of national

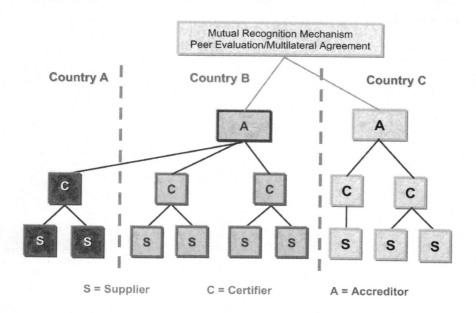

Figure 8 – Cross-border Acceptance based on Mutual Recognition between Accreditors

accreditation bodies for laboratories has established the Multilateral Recognition Arrangement (MRA) to which 84 of ILAC members from 72 countries subscribe. Members of the ILAC/MRA recognize the accreditation of CABs carried out by other members. It should be noted here, that not all members of ILAC are members of the ILAC/MRA.

In the case of management system and product certifiers, another international agreement exists that facilitates mutual recognition. This is the Multilateral Recognition Arrangement (MLA) of the International Accreditation Forum (IAF), an association of national accreditation bodies that accredit management system and product certification bodies. To date, IAF has members from 62 countries of which 54 have been admitted to at least one level of the IAF/MLA.

National accreditation bodies that practice accreditation of both laboratories and product and management system certifiers need to be members of both ILAC/MRA and IAF/MLA to have their accreditations recognized in foreign countries.

CHAPTER 6

STANDARDS AND TRADE

6.1 THE IMPACT OF TRADE GROWTH ON STANDARDS

After World War II, leading industrialized nations concluded the General Agreement on Tariffs and Trade (GATT). Other nations joined them and, through successive *rounds of multilateral trade negotiations,* a set of rules for international trade was developed. Thanks to progressive lowering of tariffs and import quotas, international trade grew steadily. The Uruguay Round resulted in the decision to establish the World Trade Organization (WTO) in 1995.

In fact, the sixty plus years following World War II witnessed a growing liberalization and a momentous growth of international trade. World export of merchandise grow 373 times from 1950 to 2012 (Figure 9), that is, at an average annual rate equal to nearly 10%! In the same period, the average annual rate of growth of World Total Product (WTP) was less than 2.3%. As a result, total world trade now represents 31.8% of WTP, a historical level never reached before.

This momentous growth of international trade had a strong impact on standards and on conformity assessment. Some of globalization trends and their impact on standards and conformity assessment are explained below:

- **The impact of the globalization of markets**
  For globalized markets to function smoothly, harmonized international standards should be applied to the products being exchanged. The harmonization of conformity assessment

Growth of World Merchandise Trade

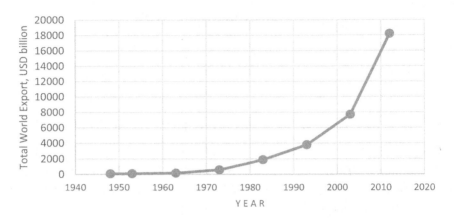

Figure 9 – Growth of World Merchandise Trade

procedures practiced by the different trading partners is the second important condition for the facilitation of international trade.

- **The impact of global industrial integration**
  Nowadays, many products are manufactured in one country using outsourced materials and components made in other countries. As long as outsourced materials are cheaper in some countries than locally produced ones, this process is bound to continue. Another common phenomenon is the delocalization of production from countries where production costs are high to countries where those costs are lower. For these processes to work properly, standards for the materials, components and products made from them should be compatible. Again, the solution is to apply compatible international standards for materials, components and final products.

- **The impact of shorter product cycles**
  The time cycle from the moment a new product in conceived and designed, materials and components identified, production processes prepared, production starts, distribution takes place until the product is phased out as more modern products replace

it, is becoming shorter and shorter. This shortening is happening due to the use of computers in design and production, the acceleration of information and communication and the acceleration of the rhythm of life in general. This situation is obliging manufacturers to use standardized parts and processes as much as possible to keep up with the fast changing environment. Developers of standards need to increase the speed of development and publication of standards as well.

- **The impact of the rising demand for higher quality**
  In today's markets, consumer have the choice of more than one product. The choice of consumers is often guided by the quality of the product. This means that manufacturers must pay more attention to quality assurance. An obvious way to achieve acceptable quality of products is to implement a quality management system according to ISO 9000 international standards. This series of standards is now being extended to more and more branches of economic activity.

- **The impact of growing environmental awareness**
  People are becoming more and more aware of the degradation of the local and global environments and the need to preserve them. This is reflected in a trend to implement standards for environmental management such as the ISO 14000 series and to certify their proper implementation.

- **The impact of the rising demand for quality and security of food**
  The rising standard of living in industrialized and developing countries is making people more aware of the importance of quality and security of foodstuffs. This is leading to a wider implementation of the international standards ISO 22000 for the quality, security and traceability (origin) of food.

- **The impact of the rising awareness of the societal responsibility of corporations**
  People are paying increasing attention to the role and responsibility of corporations in the societies where they operate.

Corporations can play a positive or negative role in the societies surrounding their operations, as a provider of employment and societal services or as a polluter. The International standard ISO 26000 provides guidance on the role played by corporations in their societal environment.

The above trends related to the rising standard of living and awareness of people around the globe are producing an important impact on standardization, conformity assessment and the related activities of testing, certification, marking and accreditation. This impact is closely related to the growth of international trade, which is making the consumption of imported products commonplace.

6.2 PRINCIPLES OF THE MULTILATERAL TRADING SYSTEM AND THEIR REFLEXION ON STANDARDS

The multilateral trading system is governed by international trade agreements concluded in the framework of the World Trade Organization (WTO). This system and agreements are based on the following principles:

- **Principle of Most Favored Nation**
  This principle states that any member of the WTO has the right to require other members to accord them a trade treatment not less favorable than what they accord to the *most favored nation*. The phrase "most favored nation" might give the first impression of unequal treatment. However, the principle implies that all members of the WTO are entitled to get EQUAL trade treatment from other members.

- **Principle of National Treatment**
  This principle states that products imported into a country should receive the same treatment as local products. This principle applies to aspects such as mandatory technical regulation, which should be the same for local and imported products.

- **Principle of Predictability and Transparency**
  Foreign companies, investors and governments should be confident that trade barriers will not be raised arbitrarily. With stability and predictability, investment is encouraged, jobs are created and consumers can fully enjoy the benefits of competition — wider choice and lower prices.

- **Principle of Reducing Trade Barriers**
  Members of the WTO continue to carry out negotiations aimed at reducing trade barriers between nations. Lowering trade barriers is one of the most obvious ways of encouraging trade. Those barriers include customs duties (or tariffs) and measures such as import bans or quotas that restrict quantities selectively.

6.3 TECHNICAL BARRIERS TO TRADE

It was recognized early on that the difference in standards and technical regulations applied in different countries represents a formidable barrier to trade. Even when zero custom duties and no quantitative restrictions (quotas) are applied, the difference in standards and technical regulations oblige manufacturers to adapt to the standards in vigor in their export markets and this represents additional work and cost.

The effect of those technical barriers was studied during the Tokyo Round (1973–1979) and by the end of that Round, the first Agreement on Technical Barriers to Trade (TBT Agreement) was prepared. The TBT Agreement was not a mandatory agreement of the GATT and, by the end of 1994, the Agreement had been signed by 46 countries and the European Union.

During the Uruguay Round (1986–1994), this Agreement was reviewed and reapproved as one of the mandatory agreements that accompanied the creation of the World Trade Organization (WTO) in 1995.

6.3.1 Principles of the TBT Agreement

The TBT Agreement is based on the following principles:

- **Technical Regulations Should not Create Unnecessary Obstacles to Trade:** The right of nations to uphold national security requirements; prevent deceptive practices; protect human health or safety, animal or plant life or health, or the environment is recognized. However, Members are required to ensure that *technical regulations are not prepared, adopted or applied **with a view to or with the effect of creating unnecessary obstacles to international trade*** (Article 2.2 of the TBT Agreement).

- **Use of International Standards as the Basis for Technical Regulations:** The Agreement stipulates that *where technical regulations are required and relevant international standards exist or their completion is imminent, Members shall use them, or the relevant parts of them, as a basis for their technical regulations* (Article 2.4).

  It is clear that, if all countries adopted international standards as the basis for technical regulations, there would be no difference between those technical regulations and no barriers to trade. However, the Agreement permits deviation from this principle, in case *international standards or relevant parts would be an ineffective or inappropriate means for the fulfilment of the legitimate objectives pursued, for instance because of fundamental climatic or geographical factors or fundamental technological problems.*

- **Full Participation in the Preparation of International Standards:** In order to harmonize technical regulations on as wide a basis as possible, the Agreement requires Members to *play a full part, within the limits of their resources, in the preparation by appropriate international standardizing bodies of international standards for products for which they either have adopted, or expect to adopt, technical regulations* (Article 2.6).

Note that the text includes the phrase "within the limits of their resources". This phrase takes into consideration the fact that developing countries may not always have sufficient human and financial resources to participate fully in the work of international standardization.

- **Technical Regulation should be Based on Performance:** The Agreement stipulates that *technical regulations should be based on performance characteristics of products and not on descriptive characteristics* (Article 2.8). This principle allows manufacturers the freedom to use the best design of products available and to improve the design without infringing the technical regulation as long as the prescribed performance is observed.

- **Notification of Technical Regulations not based on International Standards:** The Agreement stipulates that, *whenever a relevant international standard does not exist or the technical content of a proposed technical regulation is not in accordance with the technical content of relevant international standards, and if the technical regulation may have a significant effect on trade,* the Member shall *notify other Members through the Secretariat* (of the TBT Committee) *of the products to be covered by the proposed technical regulation, together with a brief indication of its objective and rationale. Such notifications shall take place at an early appropriate stage, when amendments can still be introduced and comments taken into account* (Article 2.9.2). Such an appropriate stage has been fixed by a decision of the TBT Committee to be 60 days.

  The notification article further contains the requirement that the Member proposing the technical regulation should provide to other members details and copies of the proposed technical regulation, allow enough time for making comments in writing, discuss the comments and take the comments and the results of the discussions into account.

- **Notification Requirement in Case of Urgent Problems:** The notification requirement of Article 2.9.2 may be changed in case a WTO Member encounters urgent problems of safety, health, environmental protection or national security that require the immediate adoption of a technical regulation. In such cases, the Member may proceed to the establishment and implementation of the technical regulation without allowing for the time and discussions required by the main notification Article.

Nevertheless, the Member urgently establishing a technical regulation should still *notify immediately other Members through the Secretariat of the particular technical regulation and the products covered, with a brief indication of the objective and the rationale of the technical regulation, including the nature of the urgent problems and, without discrimination, allow other Members to present their comments in writing, discuss these comments upon request, and take these written comments and the results of these discussions into account* (Article 2.10).

- **Prompt Publication of Technical Regulations and Interval before Entry into Force:** The Agreement requires members to *promptly publish adopted technical regulations* and to *allow a reasonable interval between the publication of technical regulations and their entry into force in order to allow time for producers in exporting Members, and particularly in developing country Members, to adapt their products or methods of production to the requirements of the importing Member* (Articles 2.11 and 2.12).

Concerning conformity assessment, the TBT Agreement includes specific stipulations that ensure equal treatment of imported and national products (Article 5.1.1) and that unnecessary barriers to trade are not created intentionally (Article 5.1.2).

Articles 5.2.1 and 5.2.2 of the TBT Agreement deal with the period of conformity assessment — an important aspect that can be used as a trade barrier. It is stipulated in Article 5.2.1 that conformity assessment should be *completed as expeditiously as possible and in a no less favorable order . . . than for like domestic products;* Article 5.2.2 stipulates that *the standard processing period of each conformity assessment procedure be published or that the anticipated processing period be communicated to the applicant upon request.*

Articles 5.2.3 and 5.2.4 of the Agreement deal with the information required for conformity assessment. The first of these two articles stipulates that only necessary information for conformity assessment should be requested, while the second article stipulates that information received should be kept confidential. The objective of these two articles is not to permit the divulgation of such information to other parties, such as national competitors.

Article 5.2.5 deals with the fees for conformity assessment, which should *be equitable in relation to fees for local products*, while Article 5.2.6 stipulates that the siting of conformity assessment facilities and the taking of samples should not *cause unnecessary inconvenience to applicants or their agents.*

Article 5.2.7 is intended to facilitate conformity assessment procedures in cases where the product has been modified to conform to regulations or standards. This Article stipulates that in such cases only the aspects of the product that have been modified should be retested. Article 5.2.8 requires member countries to have an appeal procedure in place *to review complaints concerning the operation of a conformity assessment procedure and to take corrective action when a complaint is justified.*

Article 5.5 requires members to *play a full part, within the limits of their resources, in the preparation of . . . international guides and recommendations for conformity assessment procedures.*

Article 5.6 deals with situations where there are no international guides or recommendations related to conformity assessment that can be followed by the members or when a WTO member decides to issue conformity assessment procedures that are not in line with international guides. In these cases, the article requires notification to be provided to other members in a similar way as in the case of technical regulations not conforming to international standards and on condition of observing similar time constraints. Article 5.7 deals with urgent situations and, as in the case of technical regulations, members are allowed not to observe the time constraints, but they are still required to immediately notify other members.

Article 5.8 requires the *prompt publication of adopted conformity assessment procedures* to ensure timely information of other members, while Article 5.9 requires members to *allow a reasonable interval between the publication of requirements concerning conformity assessment procedures and their entry into force in order to allow time for producers in exporting Members, and particularly in developing country Members, to adapt their products or methods of production to the requirements of the importing Member.*

As can be seen, the stipulations of the TBT Agreement in its different articles, are intended to set up a fool-proof system that avoids abuse of technical regulation and conformity assessment procedures by members of the WTO to protect local industries at the expense of free trade.

Together with technical regulations and conformity assessment procedures, standards are the third element that can constitute a technical barrier to trade or contribute to the establishment of such a barrier. Concerning standards, the TBT Agreement requires the standardizing bodies of members to *accept and comply with the Code of Good Practice for the Preparation, Adoption and Application of Standards of Annex 3 to the Agreement.* This Code of Good Practice sets out principles and rules such as consultation of all stakeholders, consensus and the adoption of international standards where they exist, in line with the principles and rules explained in Chapter 2 of this book. Members whose standardizing bodies accept this Code are recognized as complying with the principles of the TBT Agreement. Such acceptance is notified by the member to the ISO/IEC Information Centre in Geneva, which keeps a record of countries that have accepted the Code.

6.3.2 Implementation of Notifications — TBT Enquiry Point

In order to implement the multiple notification requirements of the TBT Agreement, the Agreement requires members to establish a TBT Enquiry Point. The function of the Enquiry Point according to Article 10 of the Agreement is *to answer all reasonable enquiries from other Members and interested parties in other Members as well as to provide the relevant documents regarding:*

- *technical regulations adopted or proposed within its territory* (Article 10.1.1);
- *standards adopted or proposed within its territory by central or local government bodies, or by regional standardizing bodies of which such bodies are members or participants* (Article 10.1.2);
- *conformity assessment procedures, or proposed conformity assessment procedures, which are operated within its territory by central or local government bodies, or by non-governmental bodies which have legal power to enforce a technical regulation, or by regional bodies of which such bodies are members or participants* (Article 10.1.3).

Additionally, the Enquiry Point should be able to answer questions and provide documents regarding *the membership and participation of the Member . . . in international and regional standardizing bodies and conformity assessment systems* (Article 10.1.4). The Enquiry Point should also indicate the place where the notifications can be obtained (Article 10.1.5). The addresses and contact details of the Enquiry Points of WTO members can be found on the site of the Organization.

The question of the languages used in notifications is evoked in the TBT Agreement. While the notifications should be provided in one of the official languages of the WTO — English, French or Spanish (Article 10.9), the Agreement requires Developed country Members, *if requested by other Members, to provide, in English, French or Spanish, translations of the documents covered by a specific notification or, in case of voluminous documents, of summaries of such documents* (Article 10.5).

6.3.3 Technical Assistance in the Framework of the TBT Agreement

Article 11 of the TBT Agreement is devoted to technical assistance. Technical assistance includes many aspects of technical regulations, standardization and conformity assessment. For example, assistance should be provided by developed to developing countries on:

- *the preparation of technical regulations* (Article 11.1)
- *the establishment of national standardizing bodies, and participation in international standardizing bodies* (Article 11.2)
- *the methods by which their technical regulations can best be met* (Article 11.3)
- *the establishment of bodies for the assessment of conformity with standards* (Article 11.4)
- *the steps that should be taken by their producers if they wish to have access to systems for conformity assessment operated by governmental or non-governmental bodies within the territory of the Member receiving the request* (Article 11.5)
- *the establishment of the institutions and legal framework which would enable them to fulfil the obligations of membership or participation in international systems* for *conformity assessment* (Article 11.6).

This last item could, for example, be applied to the establishment of accreditation bodies that would facilitate the acceptance in export markets of conformity assessment carried out in developing countries.

It should be noted that, the TBT Agreement requires that technical assistance be provided to developing countries, *if requested, and on mutually agreed terms and conditions.* This phrase is repeated in the articles of the Agreement dealing with technical assistance. It means that there is no absolute obligation to provide technical assistance without request or agreement between the donor and the receiver. It is, therefore, necessary for developing countries to request assistance and to agree with the donor(s) on the conditions under which it is provided.

6.4 SANITARY AND PHYTO-SANITARY MEASURES

Governments of all countries have the obligation to ensure that citizens are supplied with food that is safe to eat and that animal and plant health is protected. Consequently, governments implement sanitary and phyto-sanitary measures for the protection of the health of people, animals and plants. These sanitary and phytosanitary measures can take many forms, such as requiring products to come from a disease-free area, inspection of products, specific treatment or processing of products, setting allowable maximum levels of pesticide residues or limiting the permitted use of additives in food.

On the other hand, health and safety regulations can be used as an excuse for protecting domestic producers. To ensure that food offered on the market of each country, including imported food, is safe and that animal and plant health is protected and, at the same time that regulations are not used to obstruct the import of safe food and animal and plant products, an Agreement on Sanitary and Phytosanitary Measures was proposed during the Uruguay Round of the GATT. This Agreement was included in the mandatory agreements that formed the WTO on its establishment in 1995.

6.4.1 Principles of the SPS Agreement

The SPS Agreement is based on the following principles:

SPS Measures should not constitute Arbitrary Discrimination or Disguised Trade Restrictions: It is recognized, that WTO members have the right to adopt or enforce measures necessary to protect human, animal or plant life or health. However, this is subject to the requirement that *these measures are not applied in a manner, which would constitute a means of arbitrary or unjustifiable discrimination between Members where the same conditions prevail or a disguised restriction on international trade* (Articles 2.1 and 2.3 of the SPS Agreement).

SPS Measures should be Harmonized Based on International Standards and Guidelines: In order to harmonize sanitary and phytosanitary measures on as wide a basis as possible, The SPS

Agreement requires that *Members shall base their sanitary or phytosanitary measures on international standards, guidelines or recommendations, where they exist* (Article 3.1).

Full Participation in the Preparation of International Standards and Guidelines: To promote harmonization, it is necessary to develop in a consensual manner the necessary international standards and guidelines. In this respect, the SPS Agreement requires Members to *play a full part, within the limits of their resources, in the relevant international organizations and their subsidiary bodies, in particular the Codex Alimentarius Commission, the International Office of Epizootics, and the . . . International Plant Protection Convention* (Article 3.4).

Note that the SPS Agreement, unlike the TBT Agreement, actually names three international standardizing organizations. These three organizations are international *intergovernmental* organizations and not international *non-governmental* organizations like ISO and the IEC.

SPS Measures may Provide Higher Protection Based on Scientific Justification: The SPS Agreement allows Members to *introduce or maintain sanitary or phytosanitary measures, which result in a higher level of sanitary or phytosanitary protection than would be achieved by measures based on the relevant international standards, guidelines or recommendations, if there is a scientific justification* (Article 3.3).

Note: According to the SPS Agreement, **scientific justification** means an examination and evaluation by the importing member of available scientific information.

Equivalence of SPS Measures: The SPS Agreement assumes that members may accept as satisfactory SPS measures taken by other members, which may be different from their own measures, *if the exporting Member objectively demonstrates to the importing Member that its measures achieve the importing Member's appropriate level of sanitary or phytosanitary protection* (Article 4.3).

Risk Assessment – The Basis of SPS Measures: The SPS Agreement requires members to base their sanitary or phytosanitary measures on *an assessment . . . of the risks to human, animal or plant life or health, taking into account risk assessment techniques developed by the relevant international organizations* (Article 5.1).

Adaptation to Regional Conditions: The SPS Agreement requires *Members to ensure that their sanitary or phytosanitary measures are adapted to the sanitary or phytosanitary characteristics of the area — whether all of a country, part of a country, or all or parts of several countries — from which the product originated and to which the product is destined. In assessing the sanitary or phytosanitary characteristics of a region, Members shall take into account, inter alia, the level of prevalence of specific diseases or pests, the existence of eradication or control programs, and appropriate criteria or guidelines which may be developed by the relevant international organizations* (Article 6.1).

6.4.2 Notification under the SPS Agreement — SPS Enquiry Point

In order to implement the notification requirements of the SPS Agreement, the Agreement requires members to establish an Enquiry Point. The SPS enquiry point is *responsible for the provision of answers to reasonable questions as well as for the provision of relevant documents regarding:*

a. *sanitary or phytosanitary regulations adopted or proposed within its territory;*
b. *control and inspection procedures, production and quarantine treatment, pesticide tolerance and food additive approval procedures, which are operated within its territory;*
c. *risk assessment procedures, factors taken into consideration, as well as the determination of the appropriate level of sanitary or phytosanitary protection;*
d. *the membership and participation of the Member, or of relevant bodies within its territory, in international and regional sanitary and phytosanitary organizations and systems, as well as in bilateral and multilateral agreements and arrangements within the scope of the Agreement, and the texts of such agreements and arrangements* (Annex B, point 3).

6.4.3 Technical Assistance in the Framework of the SPS Agreement

Signatories of the SPS Agreement have agreed to *facilitate the provision of technical assistance to other Members, especially developing country Members, either bilaterally or through the appropriate international organizations. Such assistance may be in the areas of processing technologies, research and infrastructure, including in the establishment of national regulatory bodies* (Article 9).

The Technical assessment provision of the SPS Agreement includes the case *where substantial investments are required in order for an exporting developing country Member to fulfil the sanitary or phytosanitary requirements of an importing Member.* In this case, the developed country member *shall consider providing such technical assistance as will permit the developing country Member to maintain and expand its market access opportunities for the product involved.*

6.5   FIELDS OF APPLICATION OF THE TBT
AND SPS AGREEMENT

A frequently asked question is whether the TBT Agreement or SPS Agreement should be applied to a particular situation. The answer to this question as given in the WTO brochure "Sanitary and Phytosanitary Measures" is reproduced below.

The SPS Agreement covers all measures whose purpose is to protect:

- *human or animal health from food-borne risks*
- *human health from animal- or plant-carried diseases*
- *animals and plants from pests or diseases*
- *the territory of a country from damage caused by pests*

The TBT (Technical Barriers to Trade) Agreement covers all technical regulations, voluntary standards and the procedures to ensure that these are met, except when these are sanitary or phytosanitary measures as defined by the SPS Agreement.

Figure 10 below, extracted from the WTO Brochure indicates which of the two WTO agreements applies to a particular situation.

If the answer to any of the above questions is "yes", the SPS Agreement applies.

Is it food, drink or feed, and is its objective to protect one of those from these risks?			
Human life	Animal life	Plant life	A country
• additives, contaminants, toxins or disease-causing organisms in food or drink • plant- or animal-carried disease	• additives, • contaminants, • toxins or • disease-causing organisms in food or drink • diseases • disease-carrying organisms	• pests • diseases • disease-causing or disease-carrying organisms	• pests entering, establishing or spreading

Figure 10 – **SPS or TBT?**

The WTO brochure further gives the following examples to make the situation more clear.

Examples of the Application of the TBT and SPS Agreements

Fertilizer	Regulation on permitted fertilizer residue in food and animal feed	SPS
	Specifications to ensure fertilizer works effectively	TBT
	Specifications to protect farmers from possible harm from handling fertilizer	TBT
Food labelling	Regulation on permitted food safety: health warnings, dosage	SPS
	Regulation on size, construction/ structure, safe handling	TBT
Fruit	Regulation on treatment of imported fruit to prevent pests spreading	SPS
	Regulation on quality, grading and labelling of imported fruit	TBT
Bottled water: specifications for the bottles	Materials that can be used because safe for human health	SPS
	Permitted sizes to ensure standard volumes	TBT
Cigarette packets	Government health warning: "Smoking can seriously damage your health": the label's objective is health but it is not about food, so it is not SPS	TBT

CHAPTER 7

QUALITY AND QUALITY MANAGEMENT SYSTEMS

7.1 THE EVOLUTION OF QUALITY CONCEPTS

The growing globalization and liberalization of markets have resulted in increased competition in countries with an open economy. In fact, globalization presents both an opportunity and a challenge to local industries. The opportunity is that those industries have access to the markets of other countries. The challenge is that manufacturers from other countries also have access to their markets. The outcome is that the industries of developing countries, which have been protected for many years by tariffs or quantitative restrictions, are now under unprecedented pressure to become more competitive. This has become a condition for the survival of those industries.

Increased competitiveness can be achieved, based on three main factors: quality, price and delivery time of products and services. Other factors that affect competitiveness to a lesser extent are innovation, differentiation and flexibility in satisfying specific customer requirements.

At this point, it is important to define exactly what is meant by quality. The international standard ISO 9000 gives the following definition:

QUALITY is *"the degree to which a set of inherent characteristics of a product or service fulfils requirements"*.

Since this definition relates quality to the fulfilment of requirements, the standard further explains that requirements can be stated (for example, in a standard or a purchase document), implied (that is, something people normally take for granted) or obligatory, such as in the case where mandatory technical regulations apply. Requirements may include aspects related to:

- *performance*
- *safety*
- *usability*
- *economics*
- *dependability*
- *aesthetics*

7.2 THE EVOLUTION OF QUALITY MANAGEMENT

In traditional, pre-industrialized societies, quality was the responsibility of the craftsman. The craftsman was responsible for both the design and realization of the product. When interchangeable mass production was introduced and generalized, specialized inspection departments were created to control the quality of materials, components and final products. This resulted in a separation of the control function from the production function.

Another important development related to quality took place during World War II. Statistical methods of quality control, which had been known for some time, found wide application to support the mass production of the large quantities of arms and munitions needed for the war effort. At that time, it seemed that the best way to ensure quality was to apply statistics to the process of inspection and to the interpretation of its results.

With the end of the war and, after a rapid reconstruction period, enough products started to be produced to satisfy consumer needs and interest returned to quality as a main factor to increase sales of civilian products. A number of concepts related to quality were proposed: the *quality loop* and *integrated quality control* (known in the USA as *total quality control*).

Indeed, contrary to the practice of concentrating efforts on quality control operations carried out during production and inspection, quality science started to recognize that the creation of quality starts much before and ends much after production and inspection.

It was generally recognized, that a large number of activities that contribute to the creation of the product, its maintenance and disposal also contribute to quality. These activities start with market research and continue with product design and development, process planning and preparation. They continue further with the actual realization of the product, which includes the purchase of materials and components, the production and inspection, the packaging and storing. The activities that contribute to the creation of quality also include the activities of sales and distribution, installation and commissioning (in the case of products that need this activity), technical assistance and servicing as well as the different activities of "after sales services" (such as the provision of spare parts). In fact, it is now considered that the disposal of the product and its recycling after the end of its useful life make part of the activities that contribute to the quality of the product or service.

The activities mentioned above are conveniently divided into four categories:

1. **Determination of Customer Requirements,** including:

   • Market research
   • Contracting
   • Consideration of legal requirements

2. **Quality of Product Design**

3. **Quality of Product Conformance,** including:

   • Planning and development of production processes
   • Purchase of materials and components
   • Production or service provision
   • Verification of materials, components, semi-finished and finished products
   • Packaging the product and storing it
   • Sales and distribution activities

4. **Quality of Product Support,** including:

   • Installation and commissioning of equipment where this is done by the supplier

- After-sales services including the provision of spare parts, technical assistance or service done during or after the warranty period
- Product disposal and recycling

7.3 HOW ISO 9000 CAME INTO BEING

Based on the achievements of quality control, quality techniques started to develop in the direction of assuring the quality of products and services. Quality assurance is that part of quality management focused on providing confidence that quality requirements will be fulfilled.

Assured quality is the dream of all manufacturers. It has particular importance for certain industries such as defense and nuclear industries as well as for large scale buyers. Actually, defense departments of Western countries such as the United States and Great Britain started to develop quality assurance requirements to which they required their suppliers to conform.

At the end of the 1940s, the Department of Defense in the USA recognized the benefits of a system that had transformed the Japanese manufacturing industry. The standardized system developed by the US Department of Defense was called "quality assurance" and involved organizations establishing procedures to manage all the functions that affected the quality of manufactured products. The UK Ministry of Defense developed the General Requirements for the Assurance of Quality of Ships and Submarines or GRAQ for short. In NATO, the Allied Quality Assurance Program or AQAP was developed to assure the quality of the equipment supplied to the armed forces of NATO countries.

The British Standards Institution (BSI) decided to publish a version of AQAP as a British Standard. In 1979, British Standard BS 5750 was published in three parts dealing with quality assurance. These standards were designed to be used by any manufacturer. They enabled manufacturing enterprises to become certified, allowing them to display a mark of registration issued by the body that carried out the assessment of the quality management system. Other national standards bodies followed with national quality assurance standards such as Swiss standard SN 029100 and Canadian standard CAN Z 299.

BSI then proposed to the International Organization for Standardization (ISO) to publish an international standard based on BS 5750. ISO established a new technical committee to study this question — ISO/TC 176 Quality Management and Quality Assurance and, in 1987, the first ISO 9000 standards were published. These standards were since revised 3 times: in 1994, 2000 and 2008. The revisions of 1994 and 2000 were major revisions.

7.4 THE ISO 9000 FAMILY OF STANDARDS

The ISO 9000 Family of standards is a set of generic international standards that prescribe quality management systems applied by organizations of any type or size that

- manufacture products or components (hardware)
- manufacture software
- manufacture processed materials (such as cement or oil)
- provide services
- provide public administration functions

The main ISO 9001 standard can be applied in several situations. One is when a buyer requires the supplier in a contract to implement a quality management system according to ISO 9001. Another is when a large-scale buyer requires potential suppliers to have a quality management system, in order to be registered with the buyer as approved suppliers. A third situation exists, where a supplier decides to implement a quality management system and to have it certified by a third party certifiers as a means to attract clients and increase sales.

A fourth situation also exists, where an organization or a corporation decides to use the principles of ISO 9000 standards to improve the quality of its products and services, but is not interested in certification. In this case, the organization should implement the approaches and techniques provided in the ISO 9004 standard, which gives guidance on a wider range of objectives of a quality management system than does ISO

9001, particularly in managing for the long-term success of an organization. ISO 9004 is recommended as a guide for organizations whose top management wishes to extend the benefits of ISO 9001 in pursuit of systematic and continual improvement of the organization's overall performance. However, ISO 9004 is not intended for certification or contractual purposes.

It is worthwhile reminding the reader that, ISO 9001 prescribes *Quality System Requirements,* which are distinct from the *Technical Requirements of Product* and *Technical Requirements of Processes.* The three sets of requirements are complimentary, but neither can replace the others.

7.5 IMPLEMENTATION OF ISO 9000 IN THE WORLD

Since the first ISO 9000 standards were published in 1987, a spectacular increase in the number of organizations implementing them and obtaining certification was witnessed. By the end of 2013, over 1.1 million organizations have been certified to ISO 9001 in 170 countries (Figure 11).

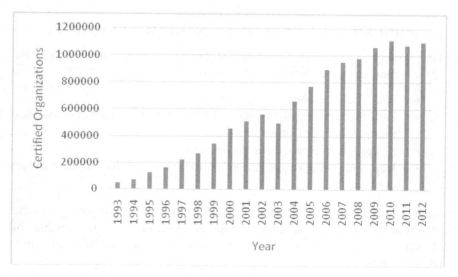

Figure 11 - ISO 9001 Implementation in the World

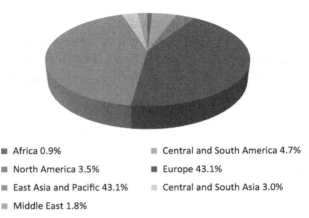

Africa 0.9%

North America 3.5%

East Asia and Pacific 43.1%

Middle East 1.8%

Central and South America 4.7%

Europe 43.1%

Central and South Asia 3.0%

Figure 12 - ISO 9001 Implementation in Regions

Unfortunately, the developing countries represent less than half of that number (Figure 12).

The following Figure 13 shows the ten countries with the highest numbers of certified organizations.

7.6 PRINCIPLES OF QUALITY MANAGEMENT SYSTEMS

The quality management system according to ISO 9001 helps the organization enhance customer satisfaction by:

- Analyzing customer requirements
- Defining processes to produce an acceptable product or service that achieves customer satisfaction
- Keeping these processes under control
- Improving the organization's processes, products and services continually

Such a quality management system provides confidence to the organization and its customers that the organization is capable of constantly supplying products or services that satisfy customer requirements.

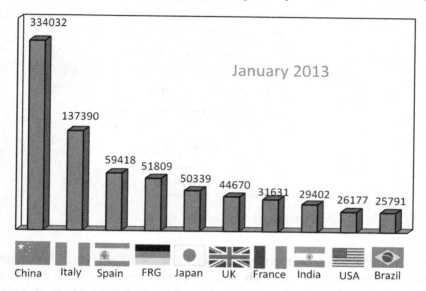

Figure 13 – The Ten Countries with the Highest Numbers of ISO 9001 Certificates

7.6.1 The Eight Quality Management Principles

The international standard ISO 9000 is based on eight quality management principles.

1. **Customer focus:** Understanding present and future customer needs in order to satisfy customer requirements and to exceed customer expectations, where possible

2. **Leadership:** Leaders establish the unity of purpose and direction of the organization. Leaders create and preserve the internal environment of an organization, so that all employees can become involved in the achievement of the organization's objectives.

3. **Involvement of people:** Full involvement of people at all levels for the benefit of the organization

4. **Process approach:** Managing resources and activities as a process allows the achievement of desired results more effectively (Figure 14).

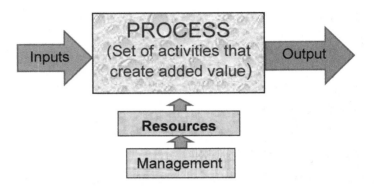

Figure 14 – The Process Approach

In a real organization many processes exist, some of which are main processes, while others are sub-processes that form a main process. The outputs of some processes are the inputs of the following processes. The processes and sub-processes may take place in different departments of the organization and they would then merge in a way that contributes to the realization of the final product or the provision of the service (Figure 15).

5. **System approach management:** Identifying, understanding and managing processes as a coherent system contributes to the realization of the objectives of the organization effectively and efficiently.

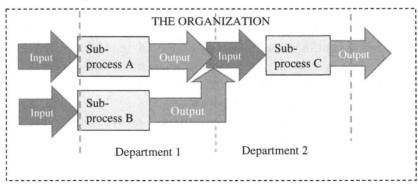

Figure 15 – Main Process, Consisting of Sub-Processes A, B and C

6. **Continual improvement:** Continual improvement should be a permanent objective of the organization.

7. **Factual approach to decision making:** Effective decisions are those based on the analysis of data and information.

8. **Mutually beneficial supplier relationship:** Such a relationship enhances the capability of the organization and its suppliers to create value.

7.6.2 Main Sections of ISO 9001

All ISO standards start with three sections entitled: scope, normative references, terms and definitions. These sections are numbered 1 to 3 and are followed by substantive sections numbered 4 and higher. The substantive sections of ISO 9001 are (Figure 16):

1. Quality management system (QMS)
2. Management responsibility
3. Resource management
4. Product realization
5. Measurement, analysis and improvement

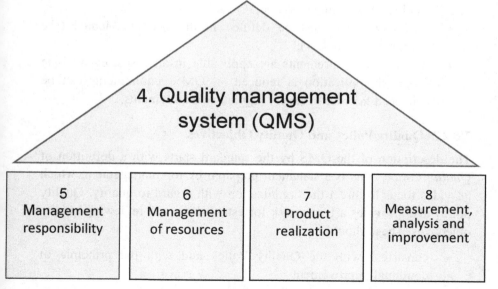

Figure 16 – Main Sections of ISO 9001

7.6.2.1 Scope of the Quality Management System according to ISO 9001

The first section of ISO 9001, Scope, stipulates that all sections of the standard should be applied by the organization implementing the standard. However, the same section allows for the application of a "reduced scope", adapted to the actual situation of the organization. Since the standard was designed to include stipulations dealing with all possible activities in an organization, some of these stipulations may turn out to be non-applicable to a certain organization due to the nature of the organization or the product or to the nature of the product realization process. For example, design control may not be applicable in an organization that produces a simple product whose design has been established long ago or is specified by the customer. A service organization may not apply control of purchased materials, if the service involves no purchased materials.

In any case, the standard stipulates that:

- The permissible exclusions should be limited to requirements to the realization of product or service and not to requirements to the Quality Management System;
- The exclusions must be defined in the Quality Manual (see section 7.6.2.4 below);
- Regulatory requirements are applicable in all cases, even if the scope of application is reduced — QMS requirements can be enhanced in order to satisfy regulatory requirements.

7.6.2.2 Quality Policy and Quality Objectives

The description of the QMS by the standard starts with a definition of *Quality Policy*. This is a statement prepared by top management, which provides focus to direct the organization with regard to quality. Quality policy also provides a framework for establishing and reviewing quality objectives. These should be

- consistent with the Quality Policy and with the principle of continual improvement
- measurable

7.6.2.3 The role of top management

ISO 9001 devotes a lot of attention to the role of top management. In Section 5. Management Responsibility, the standard enumerates the activities for which top management is responsible as follows:

- *To establish and maintain the quality policy and quality objectives and to promote them in the organization*
- *To ensure focus on customer requirements*
- *To ensure the implementation of appropriate processes to enable the achievement of quality objectives and customer requirements*
- *To establish an effective and efficient quality management system and to maintain it*
- *To ensure the availability of resources and structures to carry out necessary tasks*
- *To review the quality system periodically*
- *To decide on actions for the improvement of the quality management system*

7.6.2.4 Documentation of the quality management system — The Quality Manual

ISO 9001 insists on good documentation of the quality management system. This insistence is fully justified since documentation helps communicate intent, ensure consistency and evaluate effectiveness. It contributes to the repeatability and traceability of actions and provides objective evidence, necessary for management as well as for the auditors of the QMS.

An additional advantage of documentation is that it helps in providing training. Nevertheless, to guard against the wrong idea that documentation is needed for its own sake, ISO 9000 reminds the user that *the generation of documentation should not be an end in itself but should be a value adding activity*. The ideal quality documentation should be functional and user friendly.

The documentation of the QMS exists on a number of levels (Figure 17). The highest level is the quality policy and the quality objectives. Next, there should be a description of the processes of the quality management system, their interaction and documented procedures, that define how to carry them out.

The highest part of documentation is included in a "Quality Manual". The quality manual usually contains the quality policy, quality objective and the scope of the QMS including permissible exclusions in the product realization process. The Quality Manual may also contain the documented procedures, if their number and volume are such, that they can be included in the Manual without making it too voluminous. If there are too many procedures and their volume is too large, they may remain outside the quality manual, but the Manual should contain a reference list of them.

Other quality documents, such as the instructions for use of equipment and the quality records are necessary for quality operations. However, such documents should remain outside the Quality Manual proper, although the Manual may refer to them.

The documentation of the QMS is subject to strict control, which should be described by a mandatory documented procedure. This control is intended to ensure that QMS documents are approved for adequacy by a competent person before issue; that they are periodically reviewed, updated and reapproved and that any changes in documents and their revision status are identified. The last requirement can be easily satisfied by including in the header and the bottom of each page of documentation the document no. and date, a number indicating the revision number and its date.

The standard further requires that relevant versions of documents be available at points of use and that the documents be fully legible and identifiable. The standard gives special attention to obsolete documents, in order to avoid their getting mixed up with valid documents. Obsolete documents should be removed from points of use and, if they have to be kept for reasons of reference, they should be marked clearly to avoid unintended use. Finally, the standard requires that documents of external origin be identified and their use controlled. As expected in the age of electronic information, the standard permits the use of all forms of expression of information: paper, electronic etc.

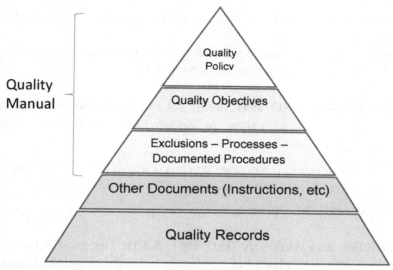

Figure 17 – Quality Management System Documentation

Quality records such as the results of tests and audits are a special kind of quality documents, which is subject to special control. This control should be specified in each organization by a documented procedure that describes how quality records are identified, stored and protected. The procedure should also indicate how to retrieve quality records, the period of their retention and the way they should be disposed of, when their retention period is past.

7.6.3 Requirements of ISO 9001 to Aspects of the Quality Management System

Quality Policy (ISO 9001, point 5.3): As mentioned in 7.6.2.3, top management should establish the quality policy of the organization, which should be an equal part of the organization's overall policies and strategy. According to the standard, *the quality policy should fulfil the following conditions:*

a) It should be appropriate to the purpose of the organization;
b) It should include a commitment to comply with customer requirements and continually improve the effectiveness of the QMS;

c) It should provide a framework for establishing and reviewing quality objectives;
d) It should be communicated and understood within the organization;
e) It should be reviewed for continuing suitability.

Quality Objectives (ISO 9001, point 5.4.1): *Top management shall ensure that quality objectives including those needed to meet requirements for the product:*

- *Are set at relevant functions and levels within the Organization*
- *Are measurable*
- *Are consistent with the Quality Policy*

Responsibility and Authority (ISO 9001, 5.5.1): The good definition of responsibility and authority of employees is a basic principle of good management. ISO 9001 requires that, *top management shall ensure that responsibilities, authorities and their interrelation are defined and communicated within the organization.*

Management Representative (ISO 9001, point 5.5.2): *Top management shall appoint a member of management who, irrespective of other responsibilities, shall have responsibility and authority that includes:*

a) *ensuring that processes needed for the QMS are established, implemented and maintained*
b) *reporting to top management on the performance of the QMS and any need for improvement*
c) *ensuring the promotion of awareness of customer requirements throughout the organization*

The role of the Management Representative is a crucial one for the QMS. This person, as spelled out by the above points of the Standard, should be the leader in all quality matters and the go-between top management and those who carry out specific quality jobs in the organization. The standard insists that the management representative should be a member of management. In fact, the higher up in the hierarchy s/he is, the better the chances of the QMS to achieve its objectives. The fact that the standard permits the management representative to have other functions beside the QMS allows this job to go to a person at a high level in the hierarchy.

Internal communication (ISO 9001, point 5.5.3): *Top management shall ensure that appropriate communication channels are established within the Organization and that communication takes place regarding the effectiveness of the QMS.* To ensure that people are involved and that potential improvement ideas are not missed, management should encourage communication and feedback from people.

Management Review (ISO 9001, point 5.6.2): To ensure the continued suitability and effectiveness of the QMS, top management needs to review the system periodically. ISO 9001 requires that:

- *Top management shall review the Organization's QMS at planned intervals, to ensure its suitability, adequacy and effectiveness;*
- *This review shall include assessing opportunities for improvement and the need for changes in the QMS including quality policy, and quality objectives;*
- *Records from the management reviews shall be maintained.*

As can be seen from the first paragraph of this requirement, no specific period is stipulated for the management review. In a mature QMS, a yearly management review is the norm; in a new QMS a shorter period is advised.

The management review should be based on objective information. The following are the sources of this information, as recommended by the standard:

a) *Results of audits*
b) *Customer feedback*
c) *Process performance and product conformity*
d) *Status of preventive and corrective action*
e) *Follow-up actions from previous management reviews*
f) *Planned changes* (in the organization) *that could affect the QMS*
g) *Recommendations for improvements*

If properly carried out, the Management Review could provide useful outputs to guide top management. Based on those outputs, management can take decisions and actions related to the improvement of the effectiveness of the QMS and its processes, improvement of the product related to customer requirements and resource needs.

Resource Management (ISO 9001, point 6): The ISO 9000 standards mention two types of resources, important for assuring the quality of products and services. These are human resources and infrastructure.

According to ISO 9001, personnel performing work affecting product quality should be competent, on the basis of appropriate:

- *education*
- *training*
- *skills*
- *experience*

To ensure the competence of personnel, the standard recommends a number of actions to be taken by the organization:

a) *determining the necessary competence for personnel performing work affecting product quality*
b) *providing training or taking other actions to satisfy these needs*
c) *evaluating the effectiveness of such actions taken*
d) *ensuring that its personnel are aware of the relevance and importance of their activities and how they contribute to the achievement of the quality objectives*
e) *maintaining appropriate records of education, training, skills and experience*

Noteworthy is the additional requirement of paragraph d), namely, *making the personnel aware of the relevance and importance of their activities and their contribution to quality objectives.* ISO 9004 contains additional recommendations to achieve the effectiveness and efficiency of the organizations' operations as well as the development and involvement of people. For example, it proposes:

- *establishing individual and team objectives*
- *involving people in objective setting and decision making*
- *recognizing and rewarding*
- *creating conditions to encourage innovation*
- *reviewing people's needs and measuring their satisfaction*

Concerning the infrastructure, which includes buildings, workspace and associated utilities, process equipment (hardware and software) and supporting services such as transport or communication, ISO 9001 requires the organization to define, provide and maintain the infrastructure by:

- *defining it in terms of function, availability, cost, safety, security and renewal*
- *developing and implementing maintenance methods (including type and frequency of maintenance and verification by element of infrastructure based on its criticality and usage)*
- *consideration of environmental issues such as pollution, waste and recycling*
- *identification and mitigation of risk associated with uncontrollable natural phenomena*

ISO 9000 standards also mention the work environment with its combination of physical and human factors. ISO 9001 requires the Organization *to determine and manage the work environment needed to achieve conformity to product requirements.* ISO 9004 adds the recommendation that *management should ensure that the work environment has a positive influence on motivation, satisfaction and performance of people.*

Product Realization (ISO 9001, Point 7): this central part of the operations of any organization includes customer-related processes, design and development, purchase of materials and components, manufacturing of the product or provision of the service, measurement and verification.

Customer-related processes start with the determination of requirements related to the product whether specified by the customer or simply necessary for use as well as legal requirements. Those requirements should be reviewed prior to the submission of the tender, acceptance of contracts or orders to ensure that:

- product requirements are defined
- contract or order requirements differing from those previously expressed are resolved
- the organization has the ability to meet the requirements

According to ISO 9001, the results of this review should be recorded (ISO 9001, point 7.2). The sources of information for the determination of requirements could be:

- requirements specified by the customer
- contract requirements
- market research
- competitor analysis
- benchmarking
- statutory requirements

To ensure that there is full understanding of customer requirements, ISO 9001 requires that *the organization ... determine and implement arrangements for:*

- *customer communication in relation to product information*
- *enquiries, contracts or order handling, including amendments*
- *customer feedback, including customer complaints*

Design and development is the activity that greatly defines the success or failure of the product or service. Customer requirements represent the most important input of the design process. Based on this input, design is carried out and produces an output – a product or service design.

ISO 9001 requires that the organization define and document:

- *functional and performance requirements*
- *regulatory requirements*
- *information derived from previous similar designs*
- *other essential requirements*

The standard further requires that these *input; be* reviewed; *incomplete, ambiguous and conflicting requirements be* resolved.

ISO 9001 requires further that design and development be checked through three processes: design validation, design verification and design review.

Design and Development Verification (ISO 9001, Point 7.3.5) is carried out to ensure that design output (which could include product specifications, requirements for training, methodology, purchasing and acceptance criteria) corresponds to design input. Verification activities include 1) comparative methods, such as alternative design and development calculations, 2) evaluation against similar products, 3) tests, simulations or trials to verify conformity to specific requirements.

Design and development validation (ISO 9001, Point 7.3.6) is performed to ensure that the product is capable of fulfilling the requirements for the intended use. Validation should be completed prior to the delivery or implementation of the product, whenever practicable. Otherwise, partial validation should be performed.

Examples of design validation are:

- validation of engineering designs prior to construction or installation
- validation of software outputs prior to installation
- validation of customer services prior to widespread introduction

Design review (ISO 9001, Point 7.3.4): Design review is conducted to evaluate the ability to fulfill requirements. It helps identify problems and propose solutions. According to the standard, *participants in such reviews shall include representatives of design and development functions and the results of the reviews shall be recorded.*

Design and development changes (ISO 9001, point 7.3.7) are a potential source of mistakes and confusion. ISO 9001 requires that these changes be identified and recorded. The standard further requires that *design and development changes be reviewed, verified, validated and* *approved before implementation. The result of these reviews should be recorded.*

Purchasing (ISO 9001, point 7.4): Purchased materials and components have a direct effect on the quality of the final product. The standard sets clear conditions for the control of purchasing processes, to ensure the conformity of purchased products. These conditions begin with the selection of suppliers. ISO 9001 requires this selection to be based on established criteria for selection, evaluation and reevaluation. The standard requires the results of evaluation and selection to be recorded.

In the beginning of the purchasing process, the organization should prepare adequate information on the purchased products, which would normally include requirements for the approval of products, but may also include requirements for processes, procedures, equipment and the qualification of personnel of the supplier. Many organizations also require that suppliers have a quality management system.

Having defined these requirements, the organization proceeds to implement inspection or other activities for verifying purchased product. The level of verification can vary according to the nature of the product and previous performance of suppliers. ISO 9001 requires that, *where the organization or its customer intends to perform verification at the supplier's premises, verification arrangements be stated in purchasing information.*

Production or Service Provision (ISO 9001, point 7.5.1): The control of production or service provision processes is the heart of quality management. The necessary conditions for realizing a conforming product are specified by ISO 9001 as follows:

- *making available work instructions*
- *using suitable equipment*
- *making available and using monitoring and measuring devices*
- *implementing monitoring and measurement*
- *implementing release, delivery and post-delivery activities*

The output of some production processes cannot be verified by subsequent monitoring or measurement. This includes processes where the deficiencies become apparent only after the product is in use, where it is impossible to verify the product, the products have a high value or processes cannot be repeated. Examples of such processes can be found in welding, in the construction of unique structures or a unique complex product.

ISO 9001, point 7.5.2, stipulates that the organization should establish arrangements for the validation of these "special" processes. Arrangements proposed by the Standard include:

- *defining criteria for review and approval of processes*
- *approval of equipment*
- *qualification of personnel*
- *use of specific methods and procedures*
- *requirements for records*
- *revalidation*

Statistical methods can help control the quality of materials and products. ISO 9001 simply mentions the possibility of such use. The following section 7.7 describes the main tools of statistical quality control.

Identification and traceability (ISO 9001, point 7.5.3): Product traceability is a requirement in some types of products (e.g. pharmaceutical products) and in some contracts. In such situations, ISO 9001 requires the organization *to identify the product by suitable means throughout product realization.* Moreover, the organization is required to *control and record the unique identification of the product.*

Since all the inspection and verification work could be useless, if the results of that work are not clearly identified, the Standard requires that the product status with respect to monitoring and measurement be identified, for example, by marking products as "controlled and conforming" or as "non-conforming" or "rejected".

Customer property (ISO 9001, point 7.5.4): Sometimes the customer gives objects or software to be used or handled by the Organization. For example:

- ingredients or components supplied for inclusion in the product
- a product supplied for repair, maintenance or upgrading
- packaging materials supplied by the customer
- customer equipment or material handled by service operations such as storage or transport
- customer intellectual property such as specifications, drawings and proprietary information

In those situations, the Standard requires that *customer property be identified, verified, protected and safeguarded*. The standard further requires, that *the organization should report to the customer, if his property is lost, damaged or found unsuitable for use*.

Preservation of the Product (ISO 9001, point 7.5.5): The preservation of the product during internal processing and delivery is a necessary condition for providing products and services of satisfactory quality. The Standard requires the organization to define and implement processes that would prevent damage, deterioration or misuse during:

- handling
- packaging
- storage
- delivery

Control of measuring and monitoring devices (ISO 9001, point 7.6): Product verification and process validation call for the implementation of measurement and monitoring. The monitoring and measurement processes to be undertaken and the monitoring and measuring devices needed are left to the discretion of the organization. The standard requires that these devices *be used and controlled in a manner, consistent with the requirements of measurement*. Moreover, the Standard requires that *measuring equipment be calibrated or verified at specified intervals, or prior to use, against standards traceable to national or international standards*. This latter requirement ensures the precision of measurements as well as their compatibility with measurements carried out in the same country and in other countries.

The Standard also requires the measuring devices to be:

- *adjusted and readjusted as necessary*
- *identified to enable calibration status to be determined* (for example, by indicating on the measurement equipment the date when the next calibration is due)
- *safeguarded from adjustments that would invalidate results*
- *protected from damage and deterioration during handling, maintenance and storage*

When the measuring equipment is found not to conform to requirements, the validity of previous measuring results should be assessed. In case computer software is used in monitoring and measurement, its ability should be confirmed.

Monitoring, Measurement, Analysis and Improvement (ISO 9001, point 8); Measurement of product or service quality and of customer satisfaction is vital for monitoring the Organization's performance. The results of measurements should be converted to information, on which it is possible to act. This information is the basis for performance improvement. ISO 9001 requires the organization to *plan and implement monitoring, measurement, analysis and improvement processes needed*:

a) *to demonstrate the conformity of the product*
b) *to ensure the conformity of the QMS*
c) *to continually improve the QMS*

The Standard adds the following mention of methods of measurement and analysis:

This shall include determination of applicable methods, including statistical techniques.

This formulation leaves the decision on whether or not to use statistical techniques to the organization based on the nature of measurement results and the suitability of such techniques for their analysis.

In Point 8.2.1 the Standard puts special emphasis on the measurement of customer satisfaction: *"the Organization shall monitor information relating to customer perception as to whether the Organization has fulfilled customer requirements. The methods for obtaining and using this information shall be determined"*. The following methods for the measurement of customer satisfaction are proposed by ISO 9004: *customer satisfaction surveys and feedback on product.*

Internal audits are an important tool to verify the performance of the QMS. The purpose of internal audits is to determine whether the QMS conforms to planned arrangements, to the requirements of ISO 9001 and to those established by the Organization; as well as to determine whether the QMS is effectively implemented and maintained. ISO 9001 requires the establishment of an *"internal audit program"* taking into consideration the status and importance of the areas to be audited and the results of previous audits.

The audit program is carried out by internal auditors selected in such a way, as to ensure the objectivity and independence of audit. This means, inter alia, that auditors of each department of the organization should come from other departments, so that auditors are not asked to audit their own work.

The standard underlines the importance of taking actions without delay to eliminate detected non-conformities and their causes. This is the responsibility of the management of the area being audited. Follow-up activities to internal audits include the verification of the actions taken and the reporting of verification results.

The responsibilities and requirements for planning and conducting internal audits and for reporting results and maintaining records should be defined in a documented procedure of the QMS (ISO 9001, point 8.2.2).

To ensure that products and services are conforming to requirements, production processes should be capable of achieving the results required for product conformity. ISO 9001 requires that the organization *apply suitable methods for monitoring production processes to demonstrate their ability to achieve planned results.* The Standard devotes attention to the verification of product characteristics at different stages (semi-finished and finished products):

Evidence of conformity with the acceptance criteria shall be maintained and records shall indicate the person(s) authorizing release of product (ISO 9001, point 8.2.4).

Dealing with non-conforming product: Unless properly identified, non-conforming product may get mixed with conforming product and be used for subsequent stages or delivered to the customer. This could have grave consequences for quality and the standard requires the organization to ensure that it does not happen. The Standard further requires that the controls and related responsibilities and authorities for dealing with nonconforming product be defined in a documented procedure of the QMS.

To enhance the measures of dealing with non-conforming product, management should empower people with authority and responsibility to report non-conformities at any stage. Indeed, non-conformities should be recorded where practical, to assist learning and provide data for improvement and analysis.

ISO 9001, point 8.3 authorizes three ways of dealing with non-conforming product:

a) Taking action to eliminate the detected non-conformity
b) Authorizing the use of non-conforming product, its release or acceptance under concession by a relevant authority and, where applicable, by the customer
c) Taking action to preclude its original intended use or application

In all these situations, records of nonconformities and subsequent action, including concessions, should be maintained. Obviously, when nonconforming product is corrected, it should be subjected to re-verification to demonstrate conformity.

The Standard also deals with the situation when non-conforming product is detected after delivery or use has started. In this case, *the organization should take action appropriate to the effects or potential effects of the nonconformity.* This action could include recall of the product for repair or replacing it with conforming product.

Continual Improvement (ISO 9001, point 8.5): Continual improvement should be a major objective of any quality conscious organization. It was included in the ISO 9001 requirements starting 1994. A number of tools are at the disposal of management to achieve continual improvement. Among them are:

- Quality policy
- Quality objectives
- Audit results
- Analysis of data
- Corrective and preventive actions
- Management review

Other tools include: empowering people to accept responsibility to identify improvement opportunities; setting objectives for people and benchmarking competitors' performance.

Involving people can be encouraged by providing recognition and rewards for achievement. Suggestion schemes can play an important role in capturing improvement opportunities. However, for these schemes to continue to produce results, management should demonstrate their interest by timely reacting to positive suggestions.

In general, there are two fundamental ways to conduct process improvement:

- Breakthrough projects which lead to revision and improvement of existing processes or implementation of new ones
- Small-step, ongoing improvement activities within existing processes

Breakthrough projects are usually carried out by cross-functional teams outside routine operations. They should be conducted using project management techniques.

People in the Organization are the best source of ideas for small-step or ongoing process improvement. People involved in improvements should be given the authority, technical support and necessary resources to carry out changes. Small-step process improvements should be controlled to understand their effects and make them a permanent feature and not an accidental event.

Corrective and preventive actions (ISO 9001, points 8.5.2 and 8.5.3): In spite of precautions, non-conformities do happen and some actions are needed to correct them.

ISO 9001 proposes the following actions to correct non-conformities:

a) reviewing non-conformities (including customer complaints)
b) determining causes of non-conformities
c) evaluating the need for action to prevent recurrence
d) determining and implementing needed action
e) recording the results of actions taken
f) reviewing corrective action taken

As can be seen from the above actions, the Standard insists, not only on correcting the non-conforming product, but also on determining the cause and taking actions to prevent recurrence of the non-conformity. As in similar situations, the Standard requires the results of actions taken to be recorded. Requirements for taking these actions should be defined in a documented procedure of the QMS.

Management should be always on the lookout for non-conformities. Sources for information on non-conformities can be found in:

- Customer complaints
- Non-conformity reports
- Internal audit reports
- Outputs from management review

- Outputs from data analysis
- Outputs from satisfaction measurements
- Other Quality Management System records
- The Organization's people
- Process measurement

It is fine to correct non-conformities, but it is still better to prevent them from occurring in the first place. This would greatly improve performance by reducing all the negative effects of non-conformities, whether on the reputation of the organization or in time and money spent to correct non-conformities. The organization should do its best to prevent non-conformities. ISO 9001 prescribes a number of preventive actions to do so:

a) determining potential nonconformities and their causes
b) evaluating the need for action to prevent occurrence of nonconformities
c) determining and implementing needed action

The Standard further requires that records of results of action taken be kept and that preventive actions taken be reviewed.

7.7 STATISTICAL QUALITY CONTROL TOOLS

7.7.1 Dealing with Variation in Product Parameters

On considering the measurement data of items such as the diameter of a bolt or the net weight of a package of butter, it is often found that, small differences exist between the values determined for different items of the same product. The differences may be so small, that measurement with a coarse instrument would not reveal them. However, on using a finer measuring instrument, the small differences become apparent.

This small variation in parameters that the manufacturer is trying hard to keep uniform is due to the presence of many small variations in the production process that cannot be completely eliminated. Examples of such variations are the variations of ambient temperature, the changes in atmospheric pressure, the change in the temperature of the lubricant used in metal cutting, the change of the temperature of tools and dies, the changes in the temperature of the measuring instruments and the force applied by them. Each one of these variations may produce a minute change in the parameters of the product but, as these variations combine in a random way, they produce a randomly variable, measurable effect on the parameters of the product.

To deal with the random small changes in the parameters of interest of products, statistical tools are needed. One of the simplest statistical tools is the histogram, explained in the following example.

Example 1: The manufacturer of 250-gram packages of butter avoids infringing trade regulations by adding a small extra weight to each package. On weighing 128 packages taken from the production line, their net weights were found to be slightly overweight. The number of packages with different values of overweight are shown in the following table.

Overweight, gram	0.5	1.0	1.5	2.0	2.5	3.0	3.5	4.0	4.5
No. of packages	2	2	8	10	16	18	23	17	13

Overweight, (cont.) gram	5.0	5.5	6.0
No. of packages	7	7	1

By representing the above data graphically, a diagram showing the distribution of the packages according to overweight is obtained (figure 18). This diagram is known as a *histogram*.

On this bell-shaped diagram one notes that, the most frequent items occur in the middle of the range at an overweight of 3.5 grams. As one

Standards & Quality

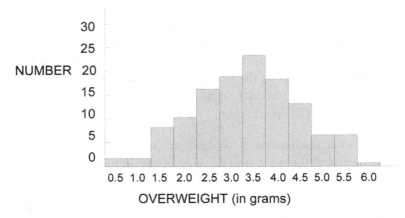

Figure 18 – Histogram of Overweight of Packages

moves along the range away from the most frequent items to the left (less overweight) or to the right (more overweight), the number of items decreases.

This typical bell shape is repeated in numerous situations and it was deduced mathematically by the German mathematician Carl Friedrich Gauss (1777 – 1855). Gauss started with just two hypotheses: 1) that the frequency decreases as one moves away from the center of the curve to the left or right and 2) that this change is symmetrical on both sides. The curve obtained by Gauss is a smooth, symmetrical one, that resembles the histogram obtained in real life, if the histogram steps on the horizontal axis were very small and the number of items was very large (in which case the number of items on the vertical axis of the diagram is replaced by their relative frequency).

The Gaussian distribution of the relative frequency of occurrence of values on a range can be found in all situations where the value is the result of the addition of many small sources of variation, which combine at random. This distribution is so commonplace that it is known as the *normal distribution*. The mathematical Gaussian distribution is perfectly symmetrical and it continues in both directions indefinitely (to plus and minus infinity). Of course, the relative frequency drops quickly to values very near to zero as one moves away from the center and it is practically considered that the curve ends at distances of ± 3σ from the centerline (Figure 19), where σ (pronounced sigma) is an important parameter of the curve, known as the *standard deviation*.

In fact, the number of items inside the limits ± 3σ is 99.73% of the total. Consequently, the number of items outside the limits ± 3σ is 0.27% on both sides, which means 0.135% on each side. In most practical production situations, it is considered that these numbers can be neglected and the curve is considered to practically end at ± 3σ. The percentage of items inside given ranges on the normal distribution curve are represented by the area under the relevant part of the curve and are shown on Figure 19.

The width of the Gaussian distribution curve depends on the standard deviation σ. The bigger σ, the wider is the curve. On the other hand, the total area under the curve represents the total of the relative frequencies of the values which is equal to 100% in all cases. Consequently, a larger σ means a wider but lower curve and vice versa. The position of the curve on the *x*-axis is defined by the central value of the studied parameter which is denoted by the Greek letter μ (pronounced miu). Figure 20 shows four Gaussian distributions with different values of σ and μ. Three of them have the same median line (μ = 0), while the fourth (light curve) has a different value of μ (= −2).

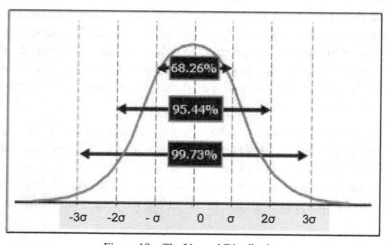

Figure 19 – The Normal Distribution

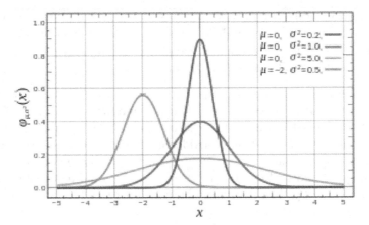

Figure 20 – Four Different Normal Distribution Curves

The acceptance limits for products are set taking into consideration the natural variation in the product parameters. Contracts often allow accepting items whose specific characteristic (weight of butter in the above example) lies above a given lower limit. This is the case where the parameter concerned is a desirable quality such as the net weight of the package of butter. Other examples of a lower limit for acceptance are the percentage of a useful ingredient and the strength of an item. In these cases, there is no need to set an upper limit, since it is to the buyer's advantage to have more of the product or to have a stronger product. For economic reasons, however, the supplier is interested in reducing the variation of his production process, to reduce the loss of material, while at the same time not going below the acceptance limit set by the contract. Figure 21 shows the relative positions of the lower limit and the distribution curve.

The acceptance limit may also be an upper limit (Figure 22). This is the case of the percentage of a harmful ingredient in a food product, which cannot be eliminated completely for economic reasons, but should be held lower than a certain limit to conform to health regulations. Other examples of an upper limit are found in the roughness of a surface and

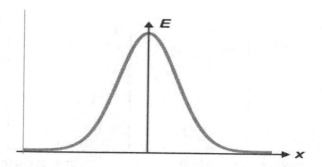

Lower limit of acceptance

Figure 21 – Distribution Curve in Relation to Lower Acceptance Limit

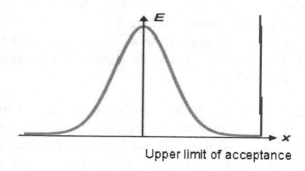

Upper limit of acceptance

Figure 22 – Distribution Curve in Relation to Upper Acceptance Limit

the out of roundness of a cylinder or a sphere, which are often required not to exceed a certain limit.

In many situations, two limits are set, between which the quality characteristic may vary. This is the case of the diameter of a bolt, which for purposes of good fitting with a hole, should lie between two limits. In this case, there is a tolerance zone limited by a lower and an upper limit. Examples of a tolerance zone limited by an upper and a lower limit are found in the sizes of metallic parts that fit together such as a piston and a cylinder and in building components, such as doors, windows, bricks, floor tiles and planks, for reasons of good fitting. A distribution curve with an upper and a lower limit is shown on Figure 23.

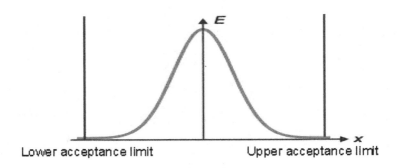

Figure 23 – Distribution Curve with Upper & Lower Acceptance Limits

In all the above cases the range of probable distribution of the quality characteristic has been shown touching the prescribed limits. For more security on delivery, the supplier can leave a security zone between the ends of the distribution curve and the limits. This, however, is often associated with higher costs.

Knowing the parameters of the distribution of items delivered by a production process (μ and σ), it is possible to estimate the expected percentage of items lying in any given range under the Gauss distribution curve. This can be done using the available tables of the relative frequency corresponding to given sections of a *standardized* Gaussian distribution. This is a Gaussian distribution with $\mu = 0$ and $\sigma = 1$. The procedure for carrying out this estimation is explained in the following example.

Example 2: A packer of coffee found that the distribution of the overweight of the packages of 500 grams follows a normal distribution with $\mu = 3.2$ g and $\sigma = 1.4$ g.

Considering that $3\sigma = 3 \times 1.4 = 4.2$ gram, the distribution curve is drawn as shown in Figure 24.

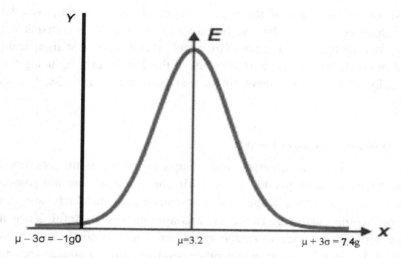

Figure 24 – Distribution Curve of Overweight of Coffee Packages

The probability that some packages have a net weight below the weight indicated on the package corresponds to the left tail of the curve shown hatched on Figure 24. In this tail the overweight is negative.

To calculate the probability of packages with negative overweight, we use the tabulated values of the standardized normal distribution. First, we have to convert the distance between the centerline of the curve and the point of interest (origin of coordinates) into a value on the standardized curve according to the formula:

$$Z = (X - \mu)/\sigma$$

Where Z = value on standardized normal distribution, X = distance of point of interest from centerline

In our case we get $Z = (0 - 3.2) / 1.4 = -2.286$ g

In the table of the standardized normal distribution (Annex A), corresponding to $Z = -2.29$ we find the value 0.01101. This means that the area under the tail of the curve (to the left of the Y-axis) represents a 1.101% probability of occurrence. Consequently, the producer faces the risk that 1.101% of his packages are under the regulatory weight. This may be considered a small enough probability that can be accepted, especially, if the regulation allows it. If the producer wishes to eliminate the risk of infringement of the regulation for all practical purposes, he should displace the centerline of the curve to the right by increasing the average overweight by 1.0 gram. This would place the whole significant part of the distribution curve to the right of the Y-axis, thus reducing the probability of negative overweight to that corresponding to 3σ, that is 0.135%.

7.7.2 Statistical Control Charts

Measuring product parameters and comparing them with prescribed values is necessary to ensure that semi-finished products are not passed to the next production stage and non-conforming products are not delivered to the customer. However, this approach is wasteful, since it detects non-conformance *after the event* and leads to costly rejects or rework. It is much better to gather process information continuously and to act immediately when deviations in the process are detected that could lead to product non-conformity.

The statistical control chart is a graphical tool for continuous monitoring of the main quality characteristic(s) of a process, to ensure that the process continues to deliver acceptable semi-finished or finished products. Control charts were first proposed by Dr. Walter Shewhart in 1924 and are sometimes known as Shewhart Control Charts.

The principle of the control chart is to carry out a simplified but continuous statistical analysis of an important process or product parameter, to plot it graphically on a time chart that would give an immediate indication when a significant change occurs in the process that can be corrected by a suitable intervention of the operator. The main idea of the control chart is to distinguish between small random variations in the process (due to the many small sources of change which are difficult and/or costly to eliminate) and more significant change which can be easily corrected. The first type of variation is considered to be "random" and "inherent in the process" and, therefore, difficult to correct without a radical remake of the process. The second type of variation is due to "assignable causes" and can be relatively easily corrected by acting on the process parameters, provided it is discovered quickly.

For example, the small random variations in the diameter of a metallic workpiece finished by grinding are those due to variation in the ambient temperature of the workshop, variation in the temperature of the workpiece, variation in the temperature of the cooling fluid, the grinding wheel and parts of the machine, variation of the sharpness of the wheel and so on. A significant change in the diameter may happen due to a slippage of the grinding wheel. Such a sudden change is significant and it can be easily corrected by resetting the wheel. Gradual variation can also happen due to wear of the grinding wheel. This unidirectional change is also assignable and it can be corrected by resetting the wheel periodically.

7.7.2.1 Control charts for variables:

Control charts can be constructed for variables, that is quantities that can take any value on a continuous scale such as the weight of a product or the length of an item, or for attributes which can take one of two discrete values such as a product being conforming or non-conforming to specifications.

The most common type of the control chart for variables is the chart for averages or $\bar{x}$ chart (read x bar chart), which is often combined with the chart for ranges or R chart. To construct this type of chart, samples are taken from the production line, each consisting of 3 to 5 items, the parameter of interest is measured for each item, which would give values denoted by x_1, x_2, x_3 etc. The average of each subgroup is calculated $\bar{x} = (x_1 + x_2 + x_3 + ...)/n$, where n is the number of items in the subgroups. The values of $\bar{x}$ are then plotted on a graph with the horizontal axis representing time.

For a process with only random sources of variation, the values of the parameter x would follow a normal distribution with a certain standard deviation, σ. Statistical science shows that the mean of several individual items $\bar{x}$ also follows a normal distribution with a smaller σ. Indeed, the standard deviation of the mean is given by the relationship:

$$\sigma_{\bar{x}} = \sigma_x / \sqrt{n}$$

where $\sigma_{\bar{x}}$ = standard deviation of the mean of sub-groups, σ_x = standard deviation of the studied quality characteristic of the produced items and n = number of items in each sub-group.

Since n is greater than 1, $\sqrt{n}$ is also greater than 1 and $\sigma_{\bar{x}}$ is less than σ_x. For example, if $n = 4$,

$$\sigma_{\bar{x}} = \sigma_x / \sqrt{4} = \sigma_x / 2$$

Having established the value of the standard deviation of the means, it is possible to construct the control chart. However, for this we need also to know the value of the central line of the chart. If there is a known desirable value of the central line, such as the center of a tolerance zone, this could be used. If no such value is given, the actual central value of the existing distribution could be used instead.

This central value of the distribution is easily obtained by taking a large number of samples k (25, say), each representing a subgroup and after calculating the averages of each sub-group (the $\bar{x}$'s), summing up all the $\bar{x}$'s and dividing the sum by the number of the samples, k.

Having calculated the $\bar{x}$ (or mean) values for the samples and the centerline of the whole distribution (denoted by $\bar{\bar{x}}$), the values of $\bar{x}$ are plotted against the time of production. A typical control chart for the mean is shown on Figure 25.

The horizontal axis represents "time" in hours or the order of drawing of the samples (since sometimes the samples are drawn only during the work shifts, which may be separated by the night pause). Points representing the values of $\bar{x}$ are plotted on the vertical axis of the chart. The points may be joined by lines to make their positions and their movement clearer.

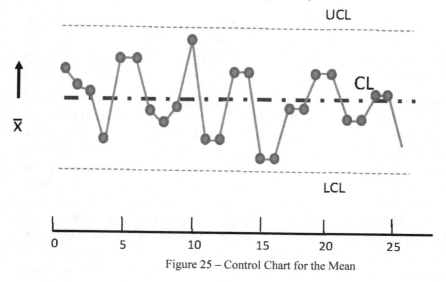

Figure 25 – Control Chart for the Mean

However, it should be remembered that, due to random variation, the actual value of the mean between two points may not be in the position indicated by the line joining them.

In addition to the Centerline (denoted by CL on the chart), two limits are drawn at a distance of $\pm 3\sigma_{\bar{x}}$ above and below the centerline. These lines are known as the Upper Control Limit and Lower Control Limit (denoted by UCL and LCL on the chart). These two limits represent statistical limits that the different points on the chart should not exceed as long as the process is under "statistical control", that is the quality characteristic (x) continues to follow a normal distribution with no shift of the central value. The typical chart shown above shows the signs of a normal distribution: there are more points nearer the centerline and less as we move away from it; the points are evenly distributed above and below the centerline and there are no trends or patterns.

When a point on the chart exceeds the UCL or LCL, this sends a signal that the centerline may have shifted upwards or downwards. It is of course possible, that a single point fell outside the limits due to the very low probability of going beyond the 3σ limits, a probability equal to 0.135%, which is quite low but not impossible. To avoid overreaction, it is recommended to wait for the next point on the chart, which may confirm or disprove the shift of the centerline. If the time between samples is considered too long or the value of the product is high, samples may be taken at shorter intervals after a point has exceeded one of the limits, to get a faster and clearer picture of what is happening in the process.

A number of tests based on probability theory exist to detect shifts in the position of the centerline even before a point crosses one of the control lines. For example, nine consecutive points falling on one side of the central line have a probability of occurrence of $(0.5)^9$, that is 0.195% which is of the same order of the 0.135% probability of crossing the three-sigma limits. One could, therefore, assume that this event indicates a shift of the centerline in the direction of the side on which the points fell. Another indicator of a shift in the central line is the occurrence of two points in a row above or below the $\pm 2\sigma$ limits. Since the probability of one point exceeding that limit on one side is 0.025, the probability of two points exceeding it in a row is 0.06% which gives a strong signal that a shift of the centerline has occurred in that direction.

If it has been determined that the centerline has indeed shifted, it is possible to return it to the desired value by an adjustment of the process: a shift of the relative position of the tool and the workpiece or a readjustment of the setting of the filling machine etc.

The control chart for $\bar{x}$ is sometimes accompanied by a control chart for the range "R". R is defined as the difference between the highest and lowest values of the quality characteristic. In other words:

$$R = x_{\max} - x_{\min}$$

where $x_{\max}$ and $x_{\min}$ are the highest and lowest values of the parameter of interest in one sub-group.

Since a number of samples (sub-groups) k, has been drawn from the production process, k values of R are available: $R_1, R_2, .. R_k$. According to statistical theory, R follows a distribution with the following parameters:

$$\text{Centerline} = \bar{R} \text{ (read } R \text{ bar)}$$
$$\text{LCL} = D_3.\bar{R}$$
$$\text{UCL} = D_4.\bar{R}$$

Where $\bar{R}$ is the mean of all values of R, D_3 and D_4 are constants dependent on the number of items n in each subgroup (the values of D_3 and D_4 are given in Annex B).

R is a measure of the inherent random variation of the process. The purpose of plotting the R chart is to follow the evolution of the variability of the process. In most practical situations, the variability is stable and the range chart seldom shows points outside the limits. However, there is an advantage in calculating the range when establishing the $\bar{x}$ chart. This advantage is due to the possibility of calculating the standard deviation of the process, σ, using the value of R rather than trying to calculate the value of the standard deviation of the process directly from the values of x. Actually, the two measure of variability are connected by the relationship:

$$\sigma = \bar{R}/d_2$$

where d_2 is a factor dependent on the number of items n in each subgroup (the values of d_2 for common n values are given in Annex 3).

Indeed, there is another, more important advantage in calculating σ from the above equation. The reason for this is that, where there is some variation in the centerline during the time of drawing the samples to construct the chart, this variation would result in an estimate of σ greater than its true value. Using this method, the effect of a possible shift (sudden or gradual) of the central line is excluded and a value of σ is obtained that corresponds better to the true variability of the process.

Usually, the R-chart is drawn below the x-chart and parallel to it. Every time that a subgroup of products is drawn, two point are plotted: one on the $\bar{x}$-chart and another on the R-chart.

In principle, it is possible to draw a control chart for individual products (x-chart). However, this is not a desirable practice. The reason is that, where the distribution of individual products is not strictly normal, the conclusions for the x-chart may be somewhat erroneous. On the other hand, statistical theory assures us that, even when the distribution of individual items is not strictly normal, the distribution of their averages is sufficiently near to a normal distribution to make conclusions based on this assumption valid for practical purposes.

7.7.2.2 Control charts for Attributes:

Attributes are qualities of the product that can take only discrete values. One type of attributes is the product being conforming or non-conforming to requirements. Another type of attributes is the number of defects in a product. Two types of control charts are constructed to follow these two types of attributes: the *p*-chart, for controlling the conformity or non-conformity of the product and the *c*-chart for controlling the number of defects in one unit (or a set quantity) of the product.

The p-chart

The percentage of non-conforming products in a given quantity of the product is denoted by "*p*". Where the probability of a product being non-conforming is constant, the value of "*p*" follows a binomial distribution.

The central line and upper and lower control limits of this distribution are given by:

$$\text{Centerline: } \bar{p} \text{ (read } p \text{ bar)}$$
$$\text{UCL: } \bar{p} + 3\sqrt{\bar{p}\,(1-\bar{p})/n}$$
$$\text{LCL: } \bar{p} - 3\sqrt{\bar{p}\,(1-\bar{p})/n}$$

Where $\bar{p}$ is the average percentage of defective (non-conforming) products and *n* is the number of products in each group. It may happen that the second term in the equation for the determination of the LCL is larger than the first term, in which case the LCL is negative. In this case, the LCL is not shown on the chart, as the concept of a negative percentage defective is meaningless.

To construct the p-chart, all items of the product produced in each shift or other suitable unit of time are tested (assuming the test to be non-destructive) and the number of defective products noted. On dividing this number by the number of items produced during the shift, the percentage defective p is established for each shift. This process is repeated for a number of shifts (25, say) and the average percentage defective is calculated as

$$\bar{p} = (p_1 + p_2 + p_3 + \ldots + p_{25})/25$$

The calculated value of $\bar{p}$ is used to construct the p-chart according to the formulas for the centerline, UCL and LCL given above. The points are then plotted on the chart whose horizontal axis represents time or the order of production. A typical *p*-chart for the percentage defective of an automatic switch is shown on Figure 26 below:

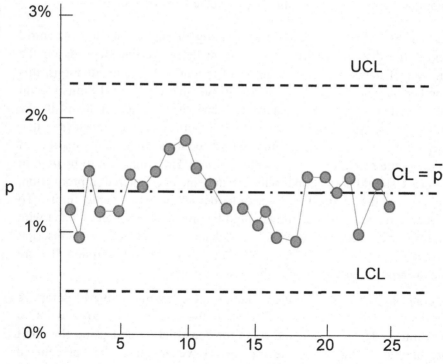

Figure 26 – Percentage Defective (*p*) Chart

After construction, the *p*-chart is used to follow-up the situation of the production (or delivery) of products. As long as the points on the chart show that the process is statistically under control, no action is needed. Action should be taken, when more than one point cross the upper control limit or when a large number of consecutive points (>13) is situated on one side of the centerline. The process then needs to be investigated in some detail, to determine and eliminate the assignable cause of the problem. It may also happen that points have a tendency to be below the centerline or fall below the lower control limit. This situation should also be investigated to discover the cause(s) of this unexpected good performance of the process (which could be the work of a more qualified employee or the use of better tools and/or equipment). If an assignable cause of this type is detected, consideration should be given to making this positive change permanent.

As can be seen from the equations fixing the upper and lower control limits, the distance of these limits from the centerline depends on the number of items in each sub-group of products. In some situations, this number may not be constant, which means that the control limits should be placed nearer to the centerline, when the number of items in the subgroup, *n*, is larger and vice-versa. Practically, it is considered that there is no need for this complicated procedure as long as the number of items in the subgroups varies within $\pm$ 10%. The value of *n* to be used in this case should correspond to the mean value encountered in production. Otherwise, the control limits should be set up as variable limits. To facilitate matters, several limits could be drawn on each chart, each with the average value of n used for its calculation indicated as a subscript. For example, UCL_{200}, UCL_{250} and UCL_{300}, where 200, 250 and 300 are the average n values to which these limits apply.

A control chart similar to the p chart is the chart of number of defective products or "*np*" chart. For this chart, there is no need to calculate a percentage defective and the rules for constructing and using the chart are similar to those for the p-chart. However, since the "percentage defective" is a more universal measure, that is readily understood by all including management, who may not be familiar with the numbers of items produced, preference often goes to the p-chart over the np-chart.

The c-chart

An important quality characteristic, which may have great significance in certain situations, is the chart for number of defects in a given quantity of the product or the c-chart. Examples of the number of defects are the number of knots in a given length of thread, the number of defects in a given area of textile material or the number of defects in a given area of tile or painted surface. The number of these defects is denoted by "c" which for small probabilities of occurrence follows a distribution known as the Poisson Distribution.

For this distribution, the centerline, upper and lower control limits are given by:

$$\text{Centerline} = \bar{c} \text{ (read c-bar)}$$

$$\text{UCL} = \bar{c} + 3\sqrt{\bar{c}}$$

$$\text{LCL} = \bar{c} - 3\sqrt{\bar{c}}$$

Where $\bar{c}$ is the mean value of c obtained by taking the average of several values of c determined when the chart is being constructed.

$$\bar{c} = (c_1 + c_2 + c_3 + \dots + c_k)/k$$

In this case, as in the case of the p-chart, a negative value of the LCL is ignored and this limit is set as the x-axis.

It should be mentioned, that in all types of control charts the first estimates of the values of $\bar{x}$, $\bar{R}$, σ, p and c give approximate values due to the limited number of the measurements made. These values could be refined after the chart has been in use for some time by making a more accurate calculation using larger numbers of measurements of those values.

7.7.3   Notions of Sampling Inspection

In many production and trade situations, it is necessary to inspect samples of items rather than their totality. A number of reasons make sampling inspection necessary. First, where the test is destructive, it is not possible to test all items. Second, testing all products of large lots may turn out to be too time consuming and costly to be carried out. The main advantage of sampling inspection is that it provides a fair idea of the quality of lots of products at a fraction of the time and cost of 100% testing. However, sampling inspection involves a certain risk of error. The application of statistics helps maximize the sureness of sampling inspection, while minimizing its cost.

As in the case of quality control charts, sampling can be carried out by attributes or by variables (see the definition of these two terms in the first paragraph of 7.7.2.1).

Sampling by Attributes:
In the case of sampling by attributes, a random sample of size N is selected from the lot of size M (where $M > N$) and tested. A fixed number Ac (usually, much smaller than N), known as the acceptance number is set and the whole lot is accepted if the number of non-conforming or defective items is less than or equal to the acceptance number Ac. If the number of defective items is greater than Ac, the whole lot is rejected.

The design of a sampling plan involves the determination of these two parameters: the sample size N and the acceptance number Ac. The choice of these two parameters depends on several factors that include the size of the lot M, the cost of the product and the cost of sampling and test operations. Chance plays an important role in sampling, since it is possible to draw more or less defective items in the sample due to the random selection of the sample items. Probability theory is, therefore, a necessary tool for the design of sampling plans and the interpretation of sampling results.

A sampling plan is characterized by its Operational Characteristic curve (OC curve). This is a curve, that shows the probability of accepting or rejecting a lot according to the actual percentage of defective items in the lot. Figure 27 shows a typical OC curve of a sampling plan.

The horizontal axis of this curve shows the actual percent defective of inspected lots; the vertical axis shows the probability that a lot with a particular percent defective is accepted based on the sampling plan. It is clear that, where there are no defects in a lot, the probability that this lot is accepted is 100% since, no matter how the sample is drawn, all items drawn pass the acceptance test. This situation is represented by the upper

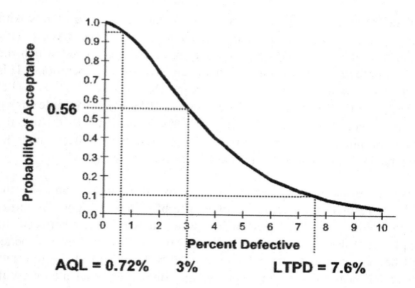

Figure 27 – Typical Operational Characteristic (OC) Curve

left-hand tip of the curve. On the other hand, a lot with a percent defective higher than zero, like the lot with 3% defective on the curve, may be accepted or rejected due to the fact that, the random selection of the items of the sample turns up more or less defective items. In the curve shown on the figure its probability of acceptance is 0.56 or 56%.

An important parameter of any sampling plan is the acceptable quality level AQL. This is the quality level of lots, that will probably be accepted 95% of the times. For the realistic buyer who has the possibility of correcting or scrapping defective items, this is a quality level at or above which his operations should still run smoothly and it is, therefore, an acceptable quality level from his point of view.

In the example of the OC curve shown, the value of AQL is 0.72%. This means that, a lot with this quality level will be accepted 95% of the times. Still, some lots with this level of defective items may be rejected 5% of the times, if by chance more defective items turn up in the sample than the acceptance number. The figure of 5% indicating the probability of rejection of a lot with the acceptable quality level is known as the "supplier's risk". This risk is due to the implementation of the sampling plan and would not exist in the case of 100% inspection.

The sampling plan may sometimes lead to the acceptance of lots with higher percent defective than the Acceptable Quality Level (AQL). This is not desirable for the buyer, who may have a limit beyond which the percent defective may cause real inconvenience for his operations. This limit is called the Lot Tolerance Percent Defective or LTPD. LTPD corresponds to a probability of acceptance of 10%. In the example shown it corresponds to a percent defective of 7.6%. This means that, using the given sampling plan, lots with this level of defective items may be accepted 10% of the time. This is known as the "buyer's risk".

It is interesting to compare the results of 100% inspection with the probable results of the sampling plan. The blue line of Figure 28 shows the probability of acceptance of lots where it was agreed between the supplier and the buyer that lots with Acceptable Quality Level or better will be accepted and lots with lower quality level will all be rejected. Under 100% inspection, this agreement would mean that all lots with percent defective equal to or lower than AQL are accepted and all lots with a higher percent defective are rejected. This gives an OC curve with squared shoulders running along the blue lines shown in Figure 28.

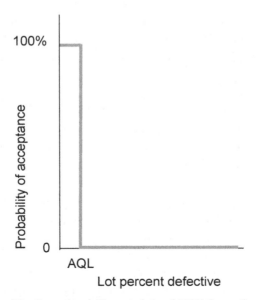

Figure 28 – Operational Charecteristic of 100% Inspection

Tables of different sampling plans with their characteristic parameters of sample size N and acceptance number Ac and the corresponding Operational Characteristic curves were established by the American statistician H. Romig in 1929. In the 1940's Romig together with H. Dodge published a classic book, which contained what became known as Dodge-Romig Sampling Inspection Tables. In 1963, the American military published a standard (Mil Std 105D) for sampling by attributes. Starting 1985, ISO published ISO 2895 on sampling by attributes in several parts.

7.8 INVESTIGATING THE CAUSES OF NON-CONFORMITY

Having carried out a statistical analysis of the quality characteristics of a product, the next logical step is to investigate the causes of non-conformity and to try to eliminate them as much as possible to improve the quality of the product or service. A useful tool in this concern is the "cause and effect" diagram, also known as the Ishikawa diagram (to honour the Japanese quality guru who proposed it) or the "fishbone diagram" because of its characteristic shape.

The principle of the cause and effect diagram is to consider the possible causes of bad quality under a number of broad headings: labor (or men), machines, methods and materials. Under each of these headings several causes may be behind insufficient quality.

Example 3: A manufacturer of furniture investigated the reasons for low quality joints in furniture. The following Ishikawa diagram (Figure 29) was established.

To ensure that all possible causes have been included in the "Cause and effect diagram", it is a good idea to carry out a **"brainstorming"** session with those who have knowledge of the process and the quality of the product or service. The participants at this session are asked to think of all possible causes of bad quality that come to their mind. All proposals

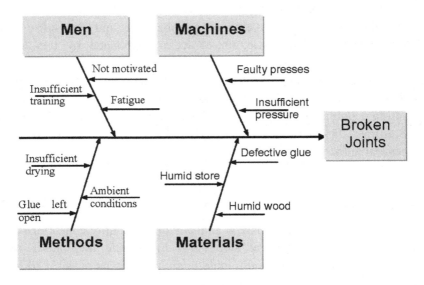

Figure 29 – Ishikawa Diagram for Broken Furniture Joints

are written on a board for all to see. It often turns out that causes that, at first, seem improbable, are indeed the real causes. It is, therefore, recommended to give full freedom to the participants of the brainstorming and not to ridicule any idea, but try to discuss all ideas with an open mind. Experience has shown that this approach often yields good results that are, in most cases, better than the results of thinking by one person.

Another useful technique for analyzing the reasons of bad quality is keeping a **"Check sheet"**. This is a sheet on which the different types of defects present in the product are recorded by the worker or the inspector. This makes the origin of defects clear, so it is easy to deal with them. The following example is the check sheet for the defects detected in a door knob.

Example 4:

<table>
<tr><td colspan="3" align="center">**Check Sheet**</td></tr>
<tr><td colspan="3">**Section:**   **Product**: Knob   **Date:**   **Inspector:**
Stage: final product   **Order no.:**   **No. of inspected units:** 142</td></tr>
<tr><td>**Defect**</td><td>**Tally**</td><td>**Number**</td></tr>
<tr><td>Surface scratches</td><td>+++ //</td><td>7</td></tr>
<tr><td>Cracks</td><td>+++ ++++//++++ +—</td><td>21</td></tr>
<tr><td>Incomplete form</td><td>++++</td><td>5</td></tr>
<tr><td>Deformed</td><td>++++ /++—</td><td>10</td></tr>
<tr><td>Other defects</td><td>///</td><td>3</td></tr>
<tr><td></td><td align="right">TOTAL</td><td>46</td></tr>
</table>

It is clear, that concentrating efforts on eliminating the most common defect (cracks in the above example) would reduce the number of defects faster than trying to eliminate the less frequent ones.

The **"Scatter diagram"** is a tool used to investigate the relationship between an observed effect and a possible cause. If in the above example of the joints in furniture, we wanted to discover to what extent the variation of the pressure of air in the pneumatic press affects the quality of the joint, we simply change the air pressure and record the percentage of defective joints corresponding to each value of air pressure. If there had been a direct relationship between the two, the points on the scatter diagram would show a clear tendency in the form of a pattern resembling a linear or other progressive (defects increasing as pressure increases) or regressive relationship (defects decreasing as the pressure increases). In figure 30 shown below, although the relation is not a neat linear relationship, it seems clear that the higher the pressure, the lower the % defectives.

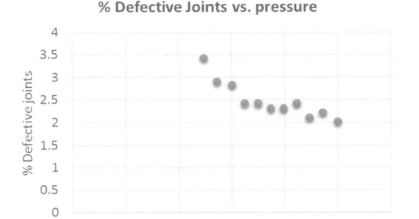

Figure 30 – Scatter Diagram for Percent Defective Joints vs. Pressure

The Pareto diagram is another effective tool to analyze defects of different types and put them in the right perspective. The frequency of occurrence of defects of different types is represented on a histogram in descending order of frequency. Then a line representing the cumulative % frequency due to the highest source alone, followed by the first two highest sources, then the sum of the three highest and so on is drawn which would eventually end at 100%. This line would typically show that a few causes of defects are behind most of the defective products, while many minor causes are behind the rest of the defective products.

In the example shown, just three causes were behind 80% of all defective products. The diagram was called after Alfredo Pareto, the Italian statistician who noticed in 1906 that about 20% of the population of a country own 80% of the land. This rule is, therefore, sometimes called the 20:80 rule.

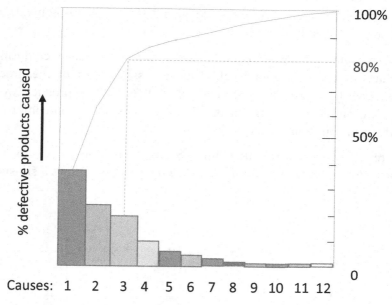

Figure 31 – Pareto Diagram

7.9 SIX-SIGMA APPROACH TO QUALITY MANAGEMENT

The six-sigma approach to quality management was introduced by Motorola in 1987 and adopted by important corporations like General Electric, Boeing, American Express and others. It represents an attempt to reduce defects to such a low level, that they are practically non-existent. In the traditional Total Quality Management approach a process with conforming products at the 3 sigma level is considered a good achievement. If a production process is so adjusted that its average is in the middle of the specification range and the points corresponding to $\pm 3\sigma$ touch the specification limits and assuming further, that the average of such a process is completely stable, the process would produce 99.73% of conforming products, which means that, 0.27% of the products may be non-conforming. If we take into consideration a possible shift of the process average towards one of the limits by one sigma, the process distribution curve would cut that limit resulting in an increase in the percentage of non-conforming products to 2.28%, which is equivalent to 228 products in a million.

The six-sigma approach requires that a process produce conforming products at the six-sigma level. With a possible shift of the process average by 1.5 sigma, this means that, 99.99966 % of the products would be conforming to requirements, leaving only 3.4 non-conforming products in a million.

Figure 32 shows how the distribution of the production process looks in relation to the specification limits in case the process follows a six-sigma approach.

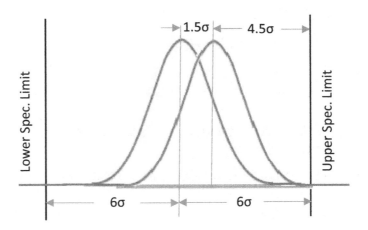

Figure 32 – Distribution of Production of Six-sigma Process

The bell-shaped normal curves represent the distribution of the process quality parameter. In the original situation (the curve to the left), the two points representing $\pm 6\sigma$ touch the specification limits. Assuming that the process average shifts to the right by 1.5σ, the whole curve moves to the right and the right hand tail of the curve now cuts the upper specification limit by that amount. This results in a number of non-conforming products corresponding to the area under the curve tail beyond 4.5σ, which is 3.4 non-conforming products per million.

A number of specific tools were developed to enable the reduction of the number of non-conforming products to the limits required by the six-sigma approach. These tools are designated by the key actions to be undertaken: **Define, Measure, Analyze, Improve** and **Control** (hence the short designation **DMAIC** – Figure 33). Below a brief explanation of these tools.

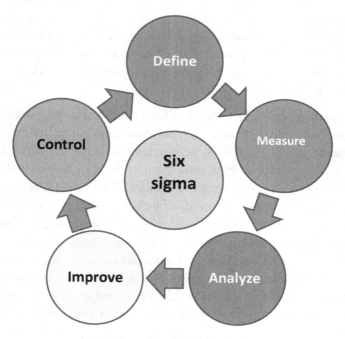

Figure 33 – Six Sigma Tools

1. **Define:** to facilitate the definition of the problem and the project. This action should be carried out by a group of experienced people from different departments of the company in the following manner.

   a) ***Establish a Process Flow chart:*** (preferably as a vertical descending line) to better visualize the process, concentrate the attention of participants on problem solving and help identify non-value-adding operations.

b) ***Brainstorm for types of non-conformance then group them in an affinity diagram***: After a brainstorming session to identify sources of non-conformance, put together the inputs provided by the members of the group under broad categories. The result will be a number of broad headings of possible sources of non-conformance with each heading filled by several such sources.

c) ***Carry out a Failure Mode & Effect Analysis (FMEA)***: Having identified all possible failure modes, assign to each potential failure mode two values (on a scale of 1-10, say) representing the **severity** and the **probability of occurrence.** Multiply the two assigned values to get a figure representing the overall gravity of each failure mode.

d) ***Establish a Value-Added Flow Chart***: Having drawn the process flow chart as a vertical descending line, indicating all operations, distinguish those operations that do not add value by a different color or by setting them aside (e.g., to the right of the flow chart). These operations include, for example, rework, set-up, inventory buffers, product movement other than customer delivery. Estimate the time spent on each operation, then sum up the total time divided into two categories: total time of essential operations and total time of non-value-adding operations. The result obtained usually shows that a lot of time is being spent unnecessarily on non-value-adding operations. This time may even exceed the time spent on essential operations. The next logical step is to try to reduce the time spent on non-value-adding operations. This can be achieved by eliminating those operations altogether, where possible, or by reducing the time spent on them drastically. The result of this action is to have a **lean** production process.

2. **Measure:** To measure the performance of the process, use suitable tools from among those described under statistical process control 7.7 above, such as:

a) ***The Pareto diagram*** to chart the main sources of non-conformance and concentrate efforts on their elimination.

b) *The histogram,* which would show whether:
- the process spread can be incorporated within the specification limits;
- the process average is properly centered between the specification limits – it is usually easy to re-center the process average by an adjustment of the tool and /or the machine or process;
- the process is behaving in a normal way (the histogram looks like a normal distribution) or is skewed.

c) *The statistical control charts* described in 7.7.2 above. By establishing one or more statistical control charts for the main quality parameters, it is possible to get a precise idea about process capability as well as trends;

It should be emphasized, that a revision of the validity of the measuring systems used should be carried out first, in order to have confidence in the figures obtained through measurements, which are used as a basis for all the above tools.

3. **Analyze:** This action is intended to get to the root causes of the problems identified and measured in the above two steps. The following tools can be used to analyze the possible causes of non-conformities, starting with those that have been identified as major ones:

a) *Ishikawa or cause and effect diagram*: to identify all possible causes of non-conformance;

b) *The scatter diagram*: which would help investigate the relationship between an effect and its possible cause(s). It should provide clear evidence that a problem is or is not caused by a particular root cause.

c) *The five-why technique*: which consists in asking several consecutive questions (about five) to really get to the root cause of a problem. This technique is based on the principle that accepting a superficial answer will not help to solve the problem.

One should, therefore, continue to question, until one gets to the real root cause, which, if remedied, the problem will disappear.

4. **Improve:** This step has the objective of finding and implementing solutions to the problems identified and, generally, identifying ways of improving the process. A number of tools can be used for this purpose:

    a) *Brainstorming*: which could be used to find solutions as well as to identify the problems.

    b) *Good housekeeping & organization of the workplace*: This certainly helps improvement since it ensures that:

        i)   unused items are not left lying around, but are sorted and removed;
        ii) workplace is orderly and things have a defined storage place;
        iii) workplace is clean;
        iv) good discipline is always kept and the above three rules are always maintained and become a standard habit applied by all employees.

        These rules of good housekeeping and workplace organization are sometimes referred to as "Japanese 5S" after the first letters of five Japanese words that signify these actions and that were popularized in 1980 by Japanese quality guru Hiroyuki Hirano.

    c) *Good reporting*: Preparing a succinct report describing the problem, the analysis carried out, the proposed solution, its implementation (when carried out) and the results thereof is essential to inform and formalize the situation. This report could fit in an A3 size sheet (or one A4 sheet recto verso). It is therefore sometimes called an A3 Report.

    d) *Corrective Action Matrix*: Since no action is effective unless it is implemented, and no action is implemented unless someone is responsible to make it happen, it is essential to keep a Corrective Action Matrix. This Matrix, which lists actions, their champions, target dates, current situation and effectiveness, is used to help keep track of who is doing what – by which deadline and with which result.

5. **Control:** Having made all the necessary corrective actions and put the process on track, it is now necessary to control the process continuously. This is carried out using the following tools:

    a) *Statistical process control* as described in 7.7 above.
    b) Establishing a *Control Plan*, which would serve as a centralized document to keep track of all significant process characteristics. Such a Plan could take the form of a table with the following information:

    - *Critical to Quality Characteristics (CTQC)*. These are characteristics known to be critical from the point of view of the client or for safety.

    - *Significant characteristics*, which are the most important parts of the CTQC. A description or reference to a standard or specification limits should be given of each.

    - *Statistical control chart(s)* for the significant characteristics with mention of its owner and location

    - An indication of the sampling frequency and measurement method including mention of the type of gauge or measuring instrument

    - A *Reaction Plan*, which would indicate the action(s) to be taken to remedy a situation of non-conformance. This Plan could be based on previous experience in dealing with problems of the same type.

7.10 COMPARISON BETWEEN ISO 9000 AND SIX-SIGMA

There is a number of differences between the ISO 9000 and six-sigma approaches to quality management. While both systems are based on a cyclical approach: in the case of ISO 9001, it is the Deming Cycle – Plan-Do-Check-Act and in the case of the six-sigma approach, it is the Define-Measure-Analyze-Improve-Control cycle, there are differences in the objectives and the degree of detail.

The ISO 9000 approach defines the actions to be taken to reach a satisfactory quality level, while the six-sigma approach has the more ambitious objective of reducing non-conforming products to a truly negligible level (3.4 in a million). On the other hand, ISO 9001 is certifiable, while six-sigma is not. Implementation of ISO 9001 and certification is recommended in the first stage of the process of quality improvement, as it includes the basic actions needed to define customer requirements and to satisfy them.

For a large corporation with a worldwide reputation, certification to ISO 9001 may not add to the reputation of the corporation and it may be a better option to start with the implementation of the six-sigma approach to make more perfect an already good quality situation. This would be the right decision, especially, where the corporation wishes to excel and to be well ahead of competitors.

CHAPTER 8

ENVIRONMENTAL MANAGEMENT SYSTEMS

8.1 MAN AND THE ENVIRONMENT — WHY ISO 14000?

Awareness of the seriousness of the environmental problems that face life on the planet has been on the rise during the last few decades. While in the 1960's people were talking mainly about local environmental problems and only a few specialists were discussing global ones, the holding of a number of international conferences with participation at the highest political level has focused attention more and more on global environmental problems that threaten to drastically and irreversibly affect the chances of continuation of life as we know it on earth.

The **Conference on the Human Environment** convened by the United Nations in Stockholm in 1972 was the first of a series of high-level conferences that focused world public attention on global environmental problems. It led to the establishment in 1984 (after affirmation by a UN General Assembly Resolution)of the **World Commission on Environment and Development** (WCED), chaired by the Prime Minister of Norway, Gro Harlem Bruntland. This Commission prepared a report published in 1987 under the title "Our Common Future", also known as the Bruntland Report, that defined the newly coined concept of "sustainable development" as "a development that meets the needs of the present without compromising the ability of future generations to meet their own needs".

A high point of international conferences on the environment was reached with the **UN Conference on Environment and Development (UNCED)**, held in Rio de Janeiro in 1992. This major Conference, also known as the **Earth Summit**, was attended by representatives of 172 countries among whom 116 heads of state. The Conference produced the

Rio Declaration, and Agenda 21, an action plan for the 21 century. Agenda 21 included 27 principles of environment and development along three main axes: poverty reduction, the production of sustainable products and the protection of the environment. It is used by countries and regions as a framework to prepare their own national Agenda 21.

A number of international conventions were opened for signature in Rio de Janeiro, which resulted in the adoption of such major conventions as the United Nations Framework Convention on Climate Change (UNFCCC), the United Nations Convention on Biological Diversity and the Statement of Forest Principles.

Recent research has revealed new risks (such as the widespread poisoning of bees by pesticides) and worse than expected outcomes for several local and global environmental problems. Badly planned and uncontrolled industrial development, as well as excessive use of agro-chemicals and the exploitation of all natural resources are impacting not only animals, plants and resources, but also the health and wellbeing of humans. Research continues to show that human activity is the main factor behind the degradation of the environment. It can be expected, that future research will reveal more and more of the negative impacts of human activities on the environment.

As environmental problems became more serious and natural resources more scarce, awareness of the need to proactively address environmental problems increased. The business world wanted to contribute more actively to efforts to improve environmental performance for a number of reasons. First, business people wanted to restore the image of industry, which was associated with pollution in the mind of the public. Second, they wanted to convey their positive intentions and actions in the field of the environment to the public. Finally, industry leaders realized that seeking better energy efficiency, recycling and economic use of materials were good for the bottom line.

During the Earth Summit, the Business Council for Sustainable Development declared that, "business and industry need tools to measure environmental performance and to develop environmental management techniques". ISO, which one year earlier had established a Strategic Advisory Group on the Environment, responded by establishing a Technical Committee, ISO/TC 207: Environmental Management to develop international standards in this field. The work of the technical committee has resulted so far in the publication of more than thirty international standards in the 14000 series dealing with environmental issues ranging from environmental management to assessment of sites and organizations, from life cycle assessment to environmental labelling and from greenhouse gases to the evaluation of carbon footprint. The following sections deal with environmental management systems.

8.2 ENVIRONMENTAL MANAGEMENT SYSTEMS

The main two ISO standards that apply to environmental management are ISO 14001 and ISO 14004. In a similar way to ISO 9001 and ISO 9004, ISO 14001 prescribes an environmental management system that can be implemented by an organization and certified, while ISO 14004 provides general guidelines on principles, systems and supporting techniques. No certification is possible to ISO 14004. Environmental management systems (EMS) are based on the same cycle as quality management systems (QMS) that is, the Plan-Do-Check-Act (PDCA) cycle. They also have the principle of continual improvement built into the management system.

8.2.1 Initial Environmental Review

The initial environmental review is an essential tool to identify the environmental situation of the organization. It is the main part of the "Plan" phase of the PDCA Cycle.

The goals of the initial environmental review are:

- To define the scope of the EMS — the environment includes many things and it is essential for the organization to have a clear vision of what it intends to manage.
- To identify *environmental aspects*, that is, ways in which the organization's activities, products and services interact with the environment and the *impacts* of those aspects

- To evaluate *aspects/impacts* in order to decide which issues require priority attention.

The initial environmental review should cover four areas:
- Legislative and regulatory requirements
- Identifying the significant impacts in the different environmental areas and ranking them according to their degree of significance
- Examination of existing environmental management practices and procedures
- Evaluation of the feedback from previous environmental incidents

In all cases, consideration should be given to normal and abnormal operation and to emergency conditions.

Key Issues that should be considered:
In order to prepare an inventory of activities, products and services of an organization susceptible of having a significant environmental impact, attention should be given to the following key issues:
- Human health: which could be impaired by carcinogenic or otherwise toxic materials, emissions harmful to health, especially, particulate materials, and other industrial emissions such as harmful or irritating noise and odor
- Ecology: where attention should be given to global warming, ozone depletion, destruction of forests, acid rain and the reduction of numbers or outright extinction of species due to the discharge of harmful gases and substances such as carbon dioxide, sulfur dioxide and heavy metals into the air, land or waterways
- Scarcity of clean fresh water resources caused by over-consumption, discharge of sewage sludge into waterways and the sea
- Depletion of fossil fuel resources such as oil and gas, which are a main source of energy and the raw material for chemicals and plastics
- Depletion of other mineral resources and some precious metals

- Waste, which could be reduced by the recovery of materials from industrial processes. Waste can also be reduced and rendered less hazardous by the composting of putrescible material and the separation and safe disposal of hazardous materials
- Land use: where severe challenges are posed by urbanization, reduction of green areas and the demand for motorways
- Transport of persons and goods: the concept of commuting over long distances to work is becoming less acceptable due to the heavy demand on polluting transport systems and roads. Trucks taking products to the market could also bring raw materials to the factory.

In estimating the significance of aspects/impacts, the organization needs to clearly define the area of control, where it has full responsibility for taking actions (aspects/impacts in that area are sometimes called direct aspects/impacts) and the area of influence, where it can only influence the actions of other players who are related to the organization (aspects/impacts in this area are known as indirect aspects/impacts). ISO 14001 requires the organization to consider aspects/impacts from both areas. However, it is obvious that the expectations in the first area are greater than in the second.

Issues in the area of control typically include:

- efficient use of raw materials
- waste minimization
- management of fresh water/waste water
- environmentally sensitive purchasing policy

The area of influence includes issues such as actions by suppliers, contractors and staff of the organization outside of its premises.

Estimation of significance takes into consideration for each aspect/impact:

- The situation vis-à-vis legal requirements (first priority)

- Environmental risk, considered as the combination of the probability of occurrence and the gravity or potential environmental damage
- Financial benefit from reduced cost of raw material, energy conservation, waste minimization
- Sensitive issues for interested parties
- Feedbacks from suppliers, clients and neighbors

One simple way of evaluating environmental risk based on a combination of probability of occurrence and gravity is to use the two-dimensional matrix shown in Figure 34 below. In this matrix, the gravity and probability of occurrence of an aspect/impact are noted with a grade from 1 to 4 on two axes and the overall environmental risk which depends on the combination of the two is defined by the resulting intersection of the two coordinates. In the figure, this risk is denoted by the number inside the corresponding rectangle (a higher number means more risk).

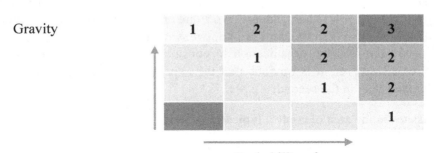

Figure 34 – Evaluation of Environmental Risk

Having carried out the evaluation of environmental risk of all significant aspects/impacts, they can be recorded by department and process in a matrix for the whole organization, which would serve as reference for further action (Figure 35 below).

Aspects/ Impacts	Ressources				Emissions						Receptors				
	Raw Material	Electrical Energy	Water	...	Air	Waste water	Solid Waste	Hazardeous waste	Soil	...	Employees(H&S)	Local community	Fauna	Flore	...
Process 1															
Sub-process 1							3		1		3				
Sub-process 2			1												
.....															
Premises															
Premise 1 : storage of hazardous material						3		3							
Premise 2 : storage of gasoil															
Premise 2 : delivery of gasoil						2									
......															
Department : design															
Modification of a production line															
Design of a new product															
...															
Department : Maintenance															
Cleansing of premises															
Maintenance of air conditionning															
....															
Department : logistics															
Transport of raw material															
Transport of the finished products															

Figure 35 - Matrix of significant aspects/impacts of the organization

8.2.2 Environmental Policy

Environmental policy is the expression of the intentions of top management to focus environmental efforts based on the outcome of the environmental review. It is the link between the environmental review and the setting of clear, detailed environmental objectives.

Obviously, the environmental policy should be "appropriate to the nature, scale and environmental impacts of the organization's activities, products and services"[g]. It should also include a commitment to comply with relevant environmental legislation and to continual improvement. The Standard ISO 14001 stipulates that environmental policy should be documented and communicated to all employees. The Standard also requires the environmental policy to be available to the public. Most interpretations of this last requirement understand it to mean "available on request" and not announced, for example, through the media.

---

[g]ISO 14001, point 4.2 (a)

8.2.3 Environmental Objectives and Targets

Having determined its significant environmental aspects/impacts and the applicable legal requirements, the organization should proceed to establishing environmental objectives based on its technological options and the available financial and operational resources. Objectives are the main directions of environmental actions. They should be specific and consistent with the policy. Targets are more detailed, could be specific to one department and should be measurable. For example, objective A may be "to reduce the use of non-biodegradable oil by 2%" and target 1 that should contribute to attaining this objective may be "to reduce the amount of oil used in process X by 10%".

Objectives and targets should be consistent with the policy, including the commitment to prevention of pollution.

8.2.4 Environmental Management Program

The next step after setting objectives and targets is to establish a program (or programs) for achieving them. The Standard requires that this program include: 1) A designation of responsibility for achieving the objectives and targets at each relevant function and level of the organization and 2) The means and time-frame by which they should be achieved.

As in other management systems, the roles, responsibilities and authorities should be defined. A management representative(s) should be appointed (who may have other responsibilities) and who should have responsibility and authority to ensure that the environmental management system is established, implemented and maintained. The management representative reports to top management on the performance of the EMS.

In addition to training, which is required for all personnel whose work may create a significant impact on the environment, the Standard gives special attention to raising awareness of personnel. Indeed, it is required that employees be made aware of the following:

a) The importance of conformance with the environmental policy and procedures and with the requirements of the EMS;
b) The significant impacts, actual or potential, of their work and the environmental benefits of improved performance;
c) Their roles and responsibilities in achieving conformance, including emergency preparedness and response and
d) The potential consequences of departure from specified operating procedures.

As with other management systems, documents should be established that describe the core elements of the management system. To avoid creating a manual of excessive size, the standard permits keeping detailed documents separately, on condition that direction to them is provided in the manual. Documents of the EMS must be controlled, which means they should be available in locations where operations essential to the effective functioning of the EMS are performed, periodically reviewed and obsolete documents removed from points of use (except where they are retained for legal purposes, in which case they must be suitably identified).

8.2.5 Environmental Communication

ISO 14001 requires that effective communication on the organization's environmental aspects/impacts be established internally and externally. However, the Standard does not require the organization to communicate externally on its significant impacts. It simply requires the organization to consider such external communication and record its decision.

8.2.6 Operational Control

Having identified significant environmental aspects, the organization should keep the activities associated with them under good operational control. The Standards requires that documented procedures stipulating operating criteria be kept where needed. The goods and services used by the organization should also be checked for environmentally significant aspects and the relevant requirements communicated to suppliers and contractors.

8.2.7 Emergency Preparedness

This is a requirement specific to environmental management systems. The organization should identify the potential for accidents and emergency situations that could have environmental impacts and be prepared to respond to them. The organization should prepare procedures for such a response, test those procedures periodically and consider their revision after the occurrence of accidents or emergency situations. This would ensure that lessons learned from such situations are not forgotten.

8.2.8 Corrective and Preventive Actions

The organization implementing ISO 14001 should monitor and measure the key characteristics of its operations that can have a significant impact on the environment using calibrated equipment. Based on this monitoring, the organization should 1) take action to mitigate any impacts caused and 2) eliminate the causes of non-conformance. In a similar way, preventive action must be taken to eliminate the causes of potential non-conformances which may occur. Records should be kept of corrective and preventive actions.

8.2.9 Environmental Management System (EMS) Audit and Management Review

The organization should carry out regular audits of the EMS to ensure that:

- The EMS conforms to planned arrangements
- The EMS has been properly implemented and maintained.

The audit program should be based on the environmental importance of the activities of the organization and on the results of previous audits.

Similarly to other management systems, the organization's top management should regularly review the EMS to ensure its continued suitability, adequacy and effectiveness. Results of the management review may be to change the policy, objectives and other elements of the EMS in the light of audit results, changing circumstances and the organization's commitment to continual improvement.

CHAPTER 9

OVERVIEW OF OTHER MANAGEMENT SYSTEMS

9.1   OTHER INTERNATIONAL STANDARDS FOR MANAGEMENT SYSTEMS

The success of the ISO 9000 series of standards on quality management (first published in 1987) and of the ISO 14000 series of standards on environmental management (first published in 1993) encouraged other users of management systems to codify them in international standards. The most popular international standards for management systems published by ISO to date are: ISO 22000 series on food safety management, ISO 26000 standard on social responsibility, ISO 27000 series on information safety management, ISO 31000 series on risk management and ISO 50000 series on energy management.

9.2   OVERVIEW OF ISO 22000 STANDARDS ON FOOD SAFETY MANAGEMENT

For obvious reasons, the safety of food is a major concern in all societies. Standards for different types of foodstuffs have been among the first standards established by most national standards bodies. With globalization and the liberalization of markets, the international trade in foodstuffs has increased several fold in the last few decades. The need for international standards that would assure the quality of foodstuffs has become more evident, in particular, to raise the confidence of consumers in the safety of imported foodstuffs.

Several concepts play a key role in food safety. **Food Safety Management Systems** *(FSMS)* is one of the key concept in this area. FSMS can be used in all stages of the food supply chain from primary production through processing, packaging and distribution. The first comprehensive food safety management system was known as Hazard

Analysis and Critical Control Points or HACCP system, which had its origins in ensuring the safety of food supplied for astronauts in NASA's first space flights. HACCP was published as a Recommended Code of Practice by the Codex Alimentarius Commission (CODEX RCP 1: 1969), which had been involved with NASA in its development. Later, HACCP was incorporated into ISO standards for food safety management.

The *traceability* of food and feed is another key concept. Traceability is crucial to allow food business operators and the authorities to withdraw or recall food products identified as unsafe. Recognizing that a supply chain is as good as its weakest link, food traceability rules require each operator to know who their immediate suppliers is and to whom they send their products.

A third key concept is *prerequisite programs (PRP)*, that is the conditions that must be established throughout the food chain and the activities and practices that must be performed in order to establish and maintain a hygienic environment.

The first of the ISO 22000 series of standards — ISO 22000 was published in 2005. The standard was prepared by ISO Technical committee 34 — "Food products" in response to the need for a standard that combines the two main requirements to the food industry: food safety and traceability.

At present, the following standards have been published in this series:
- ISO 22000: 2005–Food safety management systems–Requirements for any organization in the food chain
- ISO 22003: 2007–Food safety management systems–Requirements for bodies providing audit and certification of food safety management systems
- ISO/TS 22004: 2005–Food safety management systems–Guidance on the application of ISO 22000:2005

- ISO 22005: 2007–Traceability in the feed and food chain–
  General principles and basic requirements for system design and
  implementation

The basic standard ISO 22000:2005 sets out the requirements for a food
safety management system based on HACCP. Implementation of the
Standard can be audited and certified. It maps out what an organization
needs to do to demonstrate its ability to control food safety hazards in
order to ensure that food is safe. It can be used by any organization
regardless of its size or position in the food chain.

In the area of prerequisite programs, ISO has published one Technical
Specifications (TS) consisting of two parts:

- ISO/TS 22002-1: 2009–Prerequisite programs on food safety–
  Part 1: Food manufacturing
- ISO/TS 22002-3: 2011–Prerequisite programs on food safety–
  Part 3: Farming

9.3 STANDARDS ON SOCIAL RESPONSIBILITY

In the context of organizations, social responsibility means the
responsibility of an organization for the impact of its actions and
decisions on society and the environment. An organization demonstrates
its commitment to social responsibility both by not taking socially
harmful actions and by taking actions useful to society.

The first standard in this field was developed by Social Accountability
International (SAI) and was known as SA 8000. The standard was based
on UN and International Labor Organization (ILO) conventions and
contained nine elements:

1. Child labor
2. Forced & Compulsory Labor
3. Health & Safety
4. Freedom of Association & Right to Collective Bargaining

5. Discrimination
6. Disciplinary Practices
7. Working Hours
8. Remuneration
9. Management Systems

SA 8000 is auditable and its implementation can be certified.

In 2002, a proposal was made to ISO to develop an international standard on social responsibility. ISO Technical Management Board established a working group consisting of representatives of industry, labor organizations, employers' organizations, consumer associations, governments and NGOs. Representatives from 80 countries participated in the development of the standard, which was published in October 2010 under the title ISO 26000 — Social responsibility.

ISO 26000 is intended to assist organizations in contributing to sustainable development. It encourages them to go beyond legal compliance and to take into consideration societal, cultural, political aspects, while being consistent with international norms. ISO 26000 contains no requirements and it is, therefore, not suitable for auditing and certification.

ISO 26000 addresses seven core subjects:

1. **Organizational Governance:** every organization is responsible for its impacts on society. It should ensure transparency, respect of the law and human rights and conformity to international norms of societal behavior.
2. **Human rights:** According to the Universal Declaration of Human Rights (1948) and other international conventions that complemented it, the main human rights are:
   a. Prohibition of racial discrimination in all its forms
   b. Prohibition of discrimination against women in all its forms
   c. Prohibition of torture and other forms of undignified, cruel or inhuman treatment
   d. Children's rights: children may not participate in armed conflicts, be sold or sexually abused

    e.   Immigrant workers and their families should be protected

    f.   People should not be subjected to forced disappearance

    g.   Rights of handicapped people.

3. **Labor practices:** including:

    a.   Employment

    b.   Work conditions and social care

    c.   Societal dialogue

    d.   Safety and hygiene in the workplace

    e.   Training and human development

4. **Environmental responsibility:** including:

    a.   Prevention of pollution

    b.   Rational and sustainable use of resources

    c.   Reduction of impact on the climate and adaptation to climate change

    d.   Protection of the natural environment and the habitat of endangered species

Social Responsibility: 7 coresubjects

Figure 36 – Core Subjects Addressed by ISO 26000

5. **Fair operating practices:**
   a. Fight against corruption
   b. Responsible involvement in politics
   c. Fair competition
   d. Encouraging social responsibility in the area of influence of the organization
   e. Respect of property rights.
6. **Consumer issues:**
   a. Fulfilling the basic needs of the consumer
   b. Safety of products and services
   c. Provision of adequate and correct information
   d. Freedom of choice
   e. Listening to the consumer
   f. Correcting mistakes and compensating the consumer
   g. Raising the awareness of consumers
7. **Community involvement and development:**
   a. The organization should take the initiative to partner with the community to prevent problems and solve them, if they occur
   b. Encouragement and support of education and culture
   c. Creation of employment opportunities and development of skills of employees
   d. Increasing revenues
   e. Hygiene
   f. Investment in infrastructure

Benefits of the implementation of social responsibility principles by the organization: Organizations that implement the principles of social responsibility mentioned in ISO 26000 could achieve the following benefits:

- Improving competitiveness
- Improving the reputation of the organization
- Increasing the chances of the organization to obtain more competent employees and to attract more clients

- Improving the image of the organization vis-à-vis investors, donors, financial institutions and insurance companies
- Improving the relations between the organization and government ministries and departments, the media, other companies and the surrounding community

9.4 STANDARDS ON INFORMATION SECURITY MANAGEMENT

An organization's information system is expected to readily make available information to authorized persons, protect the confidentiality of information and ensure its integrity against change by unauthorized persons. The lack of information security may have grave consequences for the organization's business continuity, image and may have legal implications as well.

The British Standards Institution (BSI) published one of the first national standards on information security management as a code of practice (BS 7799) in the 1990's. As this code of practice matured, a second part was published (BS 7799-2), which covered information security management systems. In 2005, ISO published ISO 27001:2005–Information technology – Security techniques – Information security management systems – Requirements, which was revised in 2013. This latter version places more emphasis on how an organization measures and evaluates the performance of its Information Security Management System (ISMS). A section on outsourcing was also added in this release and additional attention was paid to the organizational context of information security.

The substantive sections of ISO 27001 deal with:

- Context of the organization
- Information security leadership
- Planning an ISMS
- Support
- Operation
- Performance evaluation
- Improvement
- Annex A – List of controls and their objectives.

The other ISO standards published in the 27000 series are:

- ISO/IEC 27002 Information technology–Security techniques–Code of practice for information security controls
- ISO/IEC 27003 Information technology–Security techniques–Information security management system implementation guidance
- ISO/IEC 27004 Information technology–Security techniques–Information security management–Measurement

9.5   ISO STANDARDS ON RISK MANAGEMENT

Given that full certainty in the outcome and continuity of situations can never be attained 100%, it is important to manage the risk due to uncertainty, so as to maximize opportunity and minimize the threat to the achievement of objectives. Risk management plays an important role in preparing an organization or a country to situations caused by market, natural or other uncertainties.

In response to the need for a global standard on risk management, ISO Technical Management Board (ISO TMB) established a working group to develop standards in this area. The WG was mandated to produce a document outlining principles and practical guidance on a risk management process applicable to all organizations, regardless of type, size, activities, location, and risk. Most importantly, the new standard was to be a guideline document, and not for certification purposes. As a result, in 2009, the WG prepared the following documents, which were published by ISO:

- ISO 31000:2009, Risk management – Principles and guidelines
- ISO Guide 73:2009, Risk management – Vocabulary

These two documents were complemented by a joint document developed by ISO and the International Electro-technical Commission:

- ISO/IEC 31010:2009, Risk management – Risk assessment techniques

9.6 ISO STANDARDS ON ENERGY MANAGEMENT

Using energy efficiently helps organizations save money as well as conserve resources and tackle climate change. ISO published the standard 50001 – Energy Management in 2011. ISO 50001 is based on the PDCA model of management systems with continual improvement.

ISO 50001:2011 provides a framework of requirements for organizations to:

- Develop a policy for more efficient use of energy
- Fix targets and objectives to meet the policy
- Use data to better understand and make decisions about energy use
- Measure the results
- Review how well the policy works
- Continually improve energy management

CHAPTER 10

THE ROLE OF METROLOGY — THE QUALITY INFRASTRUCTURE

10.1 THE APPLICATIONS OF METROLOGY

Metrology is the science of measurements. It includes all theoretical and practical aspects of measurements irrespective of the measurement uncertainty or the field of application.

Measurements are applied in all areas of human activity. They are particularly important in the following fields:

- Trade
- Industry, in particular:
  - o Manufacturing
  - o Construction
  - o Energy
  - o Communications
- Scientific research

Based on these applications, metrology is divided into three types:

- **Legal metrology**, which deals with legal measurements.
- **Industrial metrology**, which supports measurements in industry.
- **Primary (or scientific) metrology**, which keeps the national primary standards of measurement, supports scientific measurements carried out at the highest level in the country and supports legal and industrial metrology.

10.2 LEGAL METROLOGY

Many applications of metrology have a legal aspect, such as when there is a societal need to protect the buyer and seller in commercial transactions, or where measurements are used to apply sanctions.

Legislation dealing with metrology has existed in all civilized societies since the dawn of history. The following benefits are obtained thanks to metrology legislation and its practical implementation:

- Consumer protection
- Prevention of fraud
- Level playing field in trade
- Reduced costs of commercial disputes
- Full benefits from the export of commodities including raw materials and fuels
- Fair application of sanctions based on measurements (such as exceeding speed limits on roads, driving in a condition of intoxication and exceeding allowed load and/or emission levels for vehicles)
- Correct medical diagnosis and treatment based on the results of medical analyses
- Full collection of taxes (when based on measurement)
- Facilitation of trade in measuring instruments

In legal metrology the measuring instrument is designed to be used by an operator who has no special competence in metrology, the measurement procedures are simply explained in the instrument's instruction manual and the environmental conditions are not controlled, so the instrument must be designed for variable conditions. Moreover, the impartiality of the operator is not always assured, so the instrument must be protected against fraud.

Confidence in legal measurements is based on confidence in the instrument and in regulatory surveillance. Consequently, governments must establish legally binding standards for legal measuring instruments and exercise legal metrological control. Legal metrological control includes type approval of measuring instruments before they are allowed to be used in the country, their primary verification and subsequent, periodic verification and checking the observance of metrology laws and regulations, including checking the correctness of quantities of pre-packaged products and the correct use of units of measurement.

In most countries, a legal metrology authority is responsible for coordinating legal metrology activities in the country. This is usually a government body authorized to carry out certain functions of the legal metrology infrastructure and to coordinate the activities of other bodies, such as local government legal metrology services.

10.2.1 International Cooperation in Legal Metrology

In order to promote international cooperation in the field of legal metrology, the International Organization for Legal Metrology (frequently known under the French acronym OIML) was established in 1955.

The OIML is an intergovernmental organization that has published some 150 International Recommendations dealing with legal measuring instruments and a number of other documents that help members establish their national systems of legal metrological control. Examples of important OIML Recommendations are:

- OIML R76: Non-automatic weighing instruments in two parts
- OIML R111: Weights of classes E1, E2, F1, F2, M1, M1-2, M2, M2-3 and M3 in two parts
- OIML R46: Active electrical energy meters in three parts
- OIML R 49: Water meters for cold potable water and hot water in three parts

Examples of other documents published by the OIML:

- OIML D 1: Considerations for a Law on Metrology
- OIML D 14: Training and qualification of legal metrology personnel
- OIML D 16: Principles of assurance of metrological control

To avoid multiple type evaluation of legal measuring instruments before permitting their use in different countries, the OIML has implemented starting 2005 a Mutual Acceptance Arrangement (OIML MAA). This Arrangement facilitates the acceptance by national legal metrology authorities of type approval reports issued by other OIML members. Within the OIML MAA, confidence in test and examination results is reinforced by a formal and mandatory peer evaluation process.

10.3 INDUSTRIAL AND APPLIED METROLOGY

To ensure the quality of products and the proper operation of the construction, communications, energy and other economic sectors and their coordination with foreign organizations active in the same fields, it is necessary that measurements made in these sectors be carried out with the adequate precision and in conformity to international metrology standards. This implies the need to have calibration services that ensure the traceability of measurements to the national measurement standards and hence to international measurement standards.

Figure 37 shows a schematic representation of how traceability is obtained for the measuring equipment via successive calibrations. First, the working measuring equipment is calibrated against a standard, which may be kept by the factory and which is used routinely to calibrate measuring equipment. Because of its frequent use for calibrations, this standard is known as a working standard. To ensure traceability to higher level standards, the working standard is in turn calibrated against a reference standard also kept by the company, but considered their highest level standard and used infrequently to calibrate working standards. The working and reference standards of the company are kept in the measurement room or laboratory of the company under constant environmental conditions (temperature, pressure, humidity etc.).

The reference standard of the company receives its traceability thanks to calibration against the national working standard, which is used to calibrate this standard and similar standards from other companies. The national working standard is calibrated against the national standard.

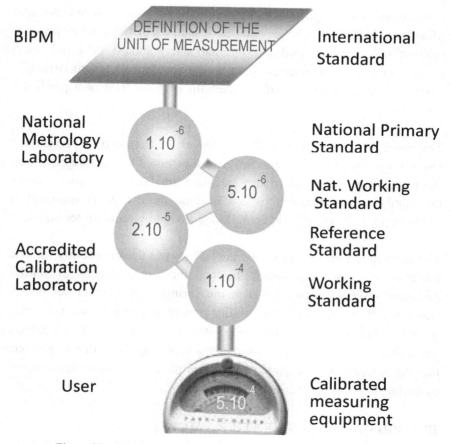

Figure 37 – Schematic View of Traceability to International Standard

Both standards are kept at the National Metrology Institute under as stable conditions as possible to guarantee their accuracy. The main difference between the two is that the national standard is used only rarely to check the national working standard, while the working standard is used routinely to calibrate reference standards of companies and other users.

To ensure the traceability of the national standard, it is compared to the international standard, if the latter is a material standard (such as the International Prototype Kilogram). Otherwise, it is checked through inter-comparisons organized by the International Bureau of Weights and Measures (BIPM) with other national standards, if there is no material international standard and the international standard is a definition relative to physical constants (such as the standard of length, which is defined as a fixed number of wavelengths of laser light of a particular frequency).

The numbers shown on the different levels of standards represent the order of uncertainty of the standard. It can be noted that as one moves up the traceability chain, uncertainty becomes smaller at the same time as the number of calibrations (which become more time consuming) is smaller and the environmental conditions are held within tighter limits.

The schematic representation of figure 37 shows the traceability for one piece of measuring equipment. In fact, there are millions of similar traceability schemes, one for each measuring instrument in the world, which together form huge traceability pyramids, one for each measured quantity (such as length, mass, temperature or pressure) with a large base consisting of the measuring instruments and the top consisting of just one material international standard or a number of realizations of the standard by national metrology institutes.

10.4 PRIMARY (OR SCIENTIFIC) METROLOGY

The upper levels of the traceability chain are assured in each country by a specialized national institute, usually known as the National Metrology Institute. National metrology institutes are established to cater to the needs of each country in providing the traceability required by the different types of applied metrology: legal, industrial as well as the needs of high-level scientific research.

The national metrology institute should be properly equipped with appropriate buildings and equipment to ensure the very high level of precision needed for realizing the national metrology standards of the country. For example, the buildings should ensure highly constant environmental conditions and freedom of shocks and vibrations. Equipment suitable for reproducing the national standards for the different types of measured quantities should be acquired, installed and maintained. The national metrology institute needs to hire and train highly qualified specialists in the different fields of metrology: length, mass, temperature etc. to operate and maintain the equipment and to liaise with other national metrology institutes to carry out the necessary inter-comparisons of standards.

10.5 THE INTERNATIONAL SYSTEM OF UNITS

Since the dawn of history, governments responded to the needs of trade by establishing standards for the important quantities measured in trade such as mass and length. The ancient Egyptians, Greeks and Romans had units of measurement prescribed by the government and represented by material measurement standards kept by the highest authorities in the land and a system of verifications of lower level measures.

For example, the ancient Egyptians used as unit of length the "Royal Cubit". This unit was equal to the length of the Pharaoh's forearm plus the width of his hand. It was represented by a Primary Standard made of black granite (Figure 38 below), one of the most dimensionally stable natural materials. The Primary Standard was kept by the "Royal Cubit Master". The working standards were wooden bars whose traceability to the Primary Standard was assured by periodic calibration every full moon.

Thanks to their well developed system of length standards the ancient Egyptians were able to achieve amazing uniformity of length measurements in the construction of the pyramids. The great pyramid of Khufu was constructed using some 2,300,000 stones, cut and polished with such accuracy that each layer was placed on top of the preceding layer without any mortar that could make up for imperfections in the size and shape of the stones. The level and squareness error in the base of the pyramid was held within 11.5 cm over a distance of 230 m, that is 0.05 %!

Figure 38 – Egyptian Papyrus showing a Two-pan Balance. A photo of the Royal Cubit made of Black Granite is superimposed below

Under weaker governments, there was a tendency to mix different systems of units and to allow different systems to prevail in the different regions of one country and the verification system was weakly enforced or completely neglected.

The French Revolution of 1789 revived the proposals of several eminent French scientists to establish a new system of units based on natural physical constants in the hope that it would be adopted by other countries. In 1790, the National Constituent Assembly adopted the principle of such a system of units and, after rejecting the idea of a unit of length based on the length of the second pendulum, due to its variability with gravity, a unit, the metre, defined as one ten-millionth of the earth's meridian was adopted.

Actual measurements of part of that meridian were made and the unit of mass was based on the adopted unit of length in such a way as to make the density of pure water a round figure (1000 kg per cubic metre). The French National Convention decided in 1995 to adopt these definitions and to establish material standards representing the units of length and mass. These were prepared and deposited in the National Archives in

1799, which is considered by some authors as the year of establishment of the decimal Metric System.

Other countries, starting with the Netherlands adopted the Metric System and in 1875 the Metre Convention was signed in Paris by 17 nations. It established a permanent organization for further developing and maintaining the metric system of units. This organization is the International Bureau of Weights and Measures known under the French abbreviation BIPM (Bureau International des Poids et Mesures). In 1889 the international prototypes for the metre and the kilogram, together with the astronomical second as unit of time, created the first international system of units.

In 1954–the Ampere, Kelvin and candela were added as base units and in 1960, the unit system was renamed as the International System of Units (SI). In 1971–the mole was added as the unit for amount of substance, bringing the total number of base units to seven (Figure 39).

The International Bureau of Weights and Measures (BIPM) is head-quartered in Sèvre near Paris, France and financed by member governments of the Metre Convention. It maintains scientific laboratories in the areas of mass, time, electricity, ionizing radiation and chemistry. The governing structure of the BIPM is formed by: (1) the General Conference on Weights and Measures (CGPM), which is made

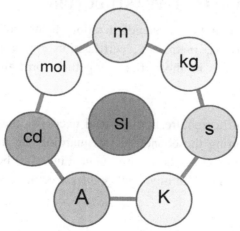

Figure 39 – Base Units of International System of Units (SI)

up of representatives of the member states and meets in Paris every four years and (2) the International Committee on Weights and Measures (CIPM), which is made up of eighteen eminent metrologists of different nationalities. The CIPM meets annually to discuss the status of international metrology and to promote worldwide uniformity of the units of measurement. The CIPM is the management board of the BIPM.

The BIPM offers high level calibrations to the national metrology institutes and is advised by Consultative Committees for the different units that bring together the highest level experts in each field of metrology.

Starting from the base units, derived units are formed that cover all areas of physical measurements. For example, starting from the unit of length, the m, the unit of area, the square metre – m^2 and the unit of volume, the cubic metre – m^3 are derived. Further, the unit of density, kilogram per cubic metre – kg/m^3 is derived. Some derived units have a name, such as the unit of pressure, which is called the Pascal and the unit of force, which is called the Newton in honour of the scientists who made great contributions to the study of phenomena to which the units are related.

10.6 THE QUALITY INFRASTRUCTURE

The national activities of standardization, technical regulation, testing, quality promotion, metrology, certification and accreditation and the institutions active in those fields together form the national Quality Infrastructure.

The quality infrastructure is a national necessity that plays an essential role in supporting the economy, safeguarding the health and safety of citizens and protecting the environment. Figure 40 below is a schematic representation of the national quality infrastructure.

The left hand part of the diagram shows the technical rule setting process, which is carried out by the national standards body (NSB) and regulators. The NSB sets voluntary national standards and the regulators in the different ministries and departments set mandatory technical regulations. These activities are linked to ISO, IEC and the Codex Alimentarius at the international level.

Next, from the left, is a schematic representation of conformity assessment activities: control of regulated products by regulators, voluntary quality mark often offered by the national standards body and management system certification performed by national standards body and other certification bodies. Under this title, also figure inspection bodies, that carry out inspection of products, services and installations.

On the right hand side, the metrology activities are shown with the national primary measurement standards at the top followed by the secondary and working standards. These standards are kept at the national metrology institute and linked to the international standards kept by the BIPM or through inter-comparisons with other national primary standards arranged by the BIPM.

Traceability is passed from the national measurement standards also to the legal metrology authority, which exercises metrological control of legal measures and measuring instruments used in the market. Traceability in industrial and other applied metrology areas is passed through calibration laboratories to companies and to test laboratories

The middle column shows accreditation activities carried out by the national accreditation body, which can accredit calibration and test laboratories as well as certification and inspection bodies. The arrows representing accreditation carry the number of the ISO standards against which accreditation of conformity assessment bodies is performed. The national accreditation body is linked to the International Laboratory Accreditation Cooperation (ILAC) and the International Accreditation Forum (IAF) at the international level.

Standards & Quality

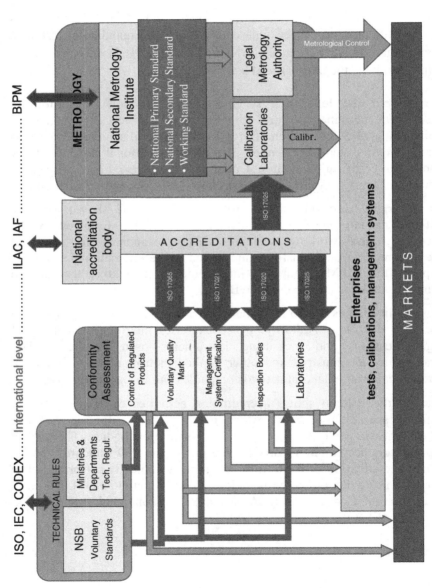

Figure 40 – Quality Infrastructure

APPENDIX A

STANDARD NORMAL DISTRIBUTION

**Table values represent
AREA to the
LEFT of the Z score
(hatched).**

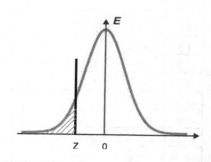

Z	.00	.01	.02	.03	.04	.05	.06	.07	.08	.09
-3.9	.00005	.00005	.00004	.00004	.00004	.00004	.00004	.00004	.00003	.00003
-3.8	.00007	.00007	.00007	.00006	.00006	.00006	.00006	.00005	.00005	.00005
-3.7	.00011	.00010	.00010	.00010	.00009	.00009	.00008	.00008	.00008	.00008
-3.6	.00016	.00015	.00015	.00014	.00014	.00013	.00013	.00012	.00012	.00011
-3.5	.00023	.00022	.00022	.00021	.00020	.00019	.00019	.00018	.00017	.00017
-3.4	.00034	.00032	.00032	.00030	.00029	.00028	.00027	.00026	.00025	.00024
-3.3	.00048	.00047	.00045	.00043	.00042	.00040	.00039	.00038	.00036	.00035
-3.2	.00069	.00066	.00064	.00062	.00060	.00058	.00056	.00054	.00052	.00050
-3.1	.00097	.00094	.00090	.00087	.00084	.00082	.00079	.00076	.00074	.00071
-3.0	.00135	.00131	.00126	.00122	.00118	.00114	.00111	.0107	.00104	.00100
-2.9	.00187	.00181	.00175	.00169	.00164	.00159	.00154	.00149	.00144	.00139
-2.8	.00256	.00248	.00240	.00233	.00226	.00219	.00212	.00205	.00199	.00193
-2.7	.00347	.00336	.00326	.00317	.00307	.00298	.00289	.00280	.00272	.00264
-2.6	.00466	.00453	.00440	.00427	.00415	.00402	.00391	.00379	.00368	.00357
-2.5	.00621	.00604	.00587	.00570	.00554	.00539	.00523	.00508	.00494	.00480
-2.4	.00820	.00798	.00776	.00755	.00734	.00714	.00695	.00676	.00657	.00639
-2.3	.01072	.01044	.01017	.00990	.00964	.00939	.00914	.00889	.00866	.00842
-2.2	.01390	.01355	.01321	.01287	.01255	.01222	.01191	.01160	.01130	.01101

Standards & Quality

-2.1	.01786	.01743	.01700	.01659	.01618	.01578	.01539	.01500	.01463	.01426
-2.0	.02275	.02222	.02169	.02118	.02068	.02018	.01970	.01923	.01876	.01831
-1.9	.02872	.02807	.02743	.02680	.02619	.02559	.02500	.02442	.02385	.02330
-1.8	.03593	.03515	.03438	.03362	.03288	.03216	.03144	.03074	.03005	.02938
-1.7	.04457	.04363	.04272	.04182	.04093	.04006	.03920	.03836	.03754	.03673
-1.6	.05480	.05370	.05262	.05155	.05050	.04947	.04846	.04746	.04648	.04551
-1.5	.06681	.06552	.06426	.06301	.06178	.06057	.05938	.05821	.05705	.05592
-1.4	.08076	.07927	.07780	.07636	.07493	.07353	.07215	.07078	.06944	.06811
-1.3	.09680	.09510	.09342	.09176	.09012	.08851	.08691	.08534	.08379	.08226
-1.2	.11507	.11314	.10123	.10935	.10749	.10565	.10383	.10204	.10027	.09853
-1.1	.13567	.13350	.13136	.12924	.12714	.12507	.12302	.12100	.11900	.11702
-1.0	.15866	.15625	.15386	.15151	.14917	.14686	.14457	.14231	.14007	.13786
-0.9	.18406	.18141	.17879	.17619	.17361	.17106	.16853	.16602	.16354	.16109
-0.8	.21186	.20897	.20611	.20327	.20045	.19766	.19489	.19215	.18943	.18673
-0.7	.24196	.23885	.23576	.23270	.22965	.22663	.22363	.22065	.21770	.22476
-0.6	.27425	.27093	.26763	.26435	.26109	.25785	.25463	.25143	.24825	.24510
-0.5	.30854	.30503	.30153	.29806	.29460	.29116	.28774	.28434	.28096	.27760
-0.4	.34458	.34090	33724	.33360	.32997	.32636	.32276	.31918	.31561	.3120
-0.3	.38209	.37828	.37448	.37070	.36693	.36317	.35942	.35569	.35197	.3482
-0.2	.42074	.41683	.41294	.40905	.40517	.40129	.39743	.39358	.38974	.3859
-0.1	.46017	.45620	.45224	.44828	.44433	.44038	.43644	.43251	.42858	.4246
-0.0	.50000	.49601	.49202	.48803	.48405	.48006	47608	.47210	.46812	.4641

APPENDIX B

CONSTANTS FOR STATISTICAL CONTROL CHARTS

n	2	3	4	5	6	7	8	9	10	11	12	13	14
d_2	1.128	1.693	2.059	2.326	2.534	2.704	2.847	2.970	3.078	3.173	3.258	3.336	3.407
D_3	0	0	0	0	0	0.076	0.136	0.184	0.223	0.256	0.283	0.307	0.328
D_4	3.267	2.574	2.282	2.114	2.004	1.924	1.864	1.816	1.777	1.744	1.717	1.693	1.673

BIBLIOGRAPHY

CEN - European Committee for Standardization (2002), Trading with and within Europe, the Benefits of standards

Cochran, C., Quality Digest (2001), Two hidden gems of continual improvement, pp. 46–49

Gerundino, D and Hilb, M, ISO Focus (June 2010), Standards: Economic and social impact, pp. 8–16

DIN - German Institute for Standardization (2000), Economic benefits of standards.

Doucet, Ch., Qualité références (2003), Continuous improvement: myth or reality? (in French)

Dusharme, D., Quality Digest (Feb 2003), Six sigma survey.

Fraser, P., Quality World, Contemplating the process.

ISO - International Organization for Standardization and IEC - International Electrotechnical Commission, ISO/IEC Guide 2: 2004 Standardization and related activities – General vocabulary

ISO - International Organization for Standardization (2013), Economic benefits of standards.

ISO - International Organization for Standardization (2013), Fast forward – National standards bodies in developing countries.

ISO - International Organization for Standardization (2009), Joining in – Participating in international standardization.

ISO - International Organization for Standardization (2010), Engaging stakeholders and building consensus.

ISO - International Organization for Standardization (2010), International standards and "private standards".

ISO - International Organization for Standardization (2012), ISO Survey

ISO - International Organization for Standardization, ISO 3:1973, amended 1998 Preferred numbers – Series of preferred numbers

ISO - International Organization for Standardization, ISO 68-1: 1998 – ISO general purpose screw threads – Basic profile, Part 1: Metric screw threads

ISO - International Organization for Standardization (1981–2003), ISO 7000 series: Graphical symbols

ISO - International Organization for Standardization, ISO 9000 series (1987–2012): Quality management systems

ISO - International Organization for Standardization, ISO 22000 series 2005–2013: Food safety management systems

ISO - International Organization for Standardization, ISO 26000: 2010 - Guidance on Social responsibility

ISO - International Organization for Standardization, ISO 50001: 2011 - Energy management systems

ISO - International Organization for Standardization and ITC – International Trade Centre (1998), ISO 9000: A workbook for service firms in developing countries

ISO - International Organization for Standardization, The Founding of ISO

Kume, H., Statistical methods for quality improvement.

Smyers, S., ISO Focus (2006), The quest for interoperability in the face of standards.

UNIDO – United Nations Industrial Development Organization and JSA (Japanese Standards Association (2001), The Pathway to excellence.

Wood, B. Quality World, Measuring up to SPC.

WTO - World Trade Organization, Agreement on Technical Barriers to Trade,

WTO- World Trade Organization, Agreement on Sanitary and Phytosanitary Measures,

INDEX

Printed in the United States
By Bookmasters